WAR, TECHNOLOGY AND THE STATE

Warren Chin

BRISTOL
UNIVERSITY
PRESS

First published in Great Britain in 2023 by

Bristol University Press
University of Bristol
1–9 Old Park Hill
Bristol
BS2 8BB
UK
t: +44 (0)117 374 6645
e: bup-info@bristol.ac.uk

Details of international sales and distribution partners are available at bristoluniversitypress.co.uk

British Library Cataloguing in Publication Data
A catalogue record for this book is available from the British Library

ISBN 978-1-5292-1340-9 hardcover
ISBN 978-1-5292-1342-3 ePub
ISBN 978-1-5292-1343-0 ePdf

Cover design: Hayes Design and Advertising
Front cover image: Epic battlefield during the postapocalypse
© user16450298 / Freepik.com
Bristol University Press use environmentally responsible print partners.
Printed and bound in Great Britain by CPI Group (UK) Ltd, Croydon, CR0 4YY

Contents

Introduction: Purpose and Scope

The debate

This book explores the relationship between technology, war and the state from the early modern period to the present and seeks to understand how technological change will impact the war–state relationship in the future. The idea for the book began when I was asked to deliver a course on the relationship between new technologies and their impact on strategy and operations at the UK's Defence Academy. This renewed interest in technology reflected the UK military's concern that a spectrum of emerging capabilities was starting to impact defence, and they were keen to understand how best to address the challenge of what they thought might be a potential military revolution. My initial position was one of profound scepticism, and I believed that technology would do no more than result in a superficial change in how the UK might engage in a future war. Indeed, the course I devised was essentially a warning from history about what happens when political leaders and their military organizations become seduced by the promises made regarding technology in war. To this end, a range of case studies, from Hitler's insane pursuit of wonder weapons in the Second World War to the emasculation of the United States' conceived revolution in military affairs in Afghanistan and Iraq, demonstrated why technology is no substitute for an understanding of the geographical and political context of war (Kaplan, 2012). However, my research on the subject challenged the assumptions that shaped this course. More importantly, it also seemed to test the conventional view that prevails in broader academic debates on the relative importance of technology as a driver of change in the domain of war. The fundamental purpose of this book is to ask if we have arrived at a point where current patterns of technological change require a reassessment of the relationship between technology, war and politics, expressed here in the form of the state.

A cursory examination of the literature on this subject of war and technology reveals what can be described as an orthodox school and a revisionist school. The traditional view of technology asserts it has played a positive role in human history, and its impact on the conduct of war resulted in the emergence of a modern military system which fits into the broader aspiration of contemporary society. This school consists of three subsets. The first is a range of narrowly focused discussions on weapons and their capability (van Creveld, 1989; Ferguson et al, 2017).

Second is the research on how military technologies and the tactical paradigms they created transformed war. The study of German and Russian operational techniques in the interwar period and how they were used in the Second World War provides the best example of this literature (Murray and Millett, 1996; Glantz, 2012; Glantz, 2015). We might also include the vast literature on the impact of nuclear weapons on the conduct of war (Freedman, 2003). It is also vital to recognize similar studies have been conducted on more recent wars, including the 1991 Gulf War (Cohen, 1996) and the campaigns waged during the early days of the war on terror. Sometimes military historians linked these new tactical paradigms together to present a history of warfare based mainly on a series of technologically based revolutions (Krepinevich, 1994; Lind, 1989). In these cases, the focus of the study remained firmly fixed on the utility of these weapons on the battlefield.

Third, and most important in this study, is the research on how military technology and the emergence of new tactical paradigms transformed not just the domain of war but also how this fed more widely into changes in society and politics. Perhaps the best example of this connection between the battlefield and changes in politics and culture is Michael Roberts' thesis on warfare in early modern Europe and the emergence of the modern state (1995). This thesis has played an instrumental role in shaping the debates on the relationship between war and the nation state. Michael Mann's three-volume study of the sources of social power (1986) and Charles Tilly's more explicit work on the relationship between war and the state (1992) are excellent examples of how war shapes politics.

Finally, there are those studies that adopt a more expansive vista when looking at how technology changes war. Perhaps the best-known book of this kind was the Tofflers' tome *War and Anti-War* (1993), which adopted a crude Marxist logic and demonstrated how technology led to the creation of three distinct modes of production, which in turn, led to the creation of different forms of warfare. The critical point is that the chain of causation is more convoluted and external to the military domain rather than emerging from within.

The revisionist school challenges the central role assigned to technology as an agent of change within war and its broader economic, political, and social connections. For example, while there is a consensus that war gave rise to the emergence of the state the reasons for this are contested. This is perhaps why Tilly, one of the most influential commentators on the war–state relationship, was sceptical about the importance of technology in this process and focused instead on the economics of waging war (Tilly, 1985). I have already acknowledged this is a perfectly reasonable position to take (Chin, 2019). The history of war is characterized by long phases of technological stagnation punctuated by occasional spasms of revolutionary change caused by various forces (Knox and Murray, 2001).

This school emphasizes the importance of political and social changes in shaping the evolution of war. Black (1994), for example, recognizes the importance of the expanding bureaucratic apparatus in states like France and Prussia in the 17th century. He believes this drive for efficiency was not so much a consequence of technological change but stemmed from the desire to increase the size of national armies and navies, which were primarily political choices dictated by monarchs' eager to pursue hegemony or prevent it (Black, 1994, 10).

Parrott (2012) also challenges the thesis of war leading to the creation of the modern state in Europe. He points to the importance of religious reformation and the drive by rulers to impose norms and beliefs on their populations that resulted in the creation of distinctive national cultures. He also emphasizes the role of economics in precipitating a shift in state power. Of significance here was the combined effect of rapid population growth across Europe, the impact of inflation on living standards, increasing concentrations of wealth and hence rising inequality and the rising social and political tensions this caused. He notes that a critical response to these trends was for ruling elites to lean towards more authoritarian forms of governance symbolized by a powerful state (Parrot, 2012, 12).

The evolution of this debate saw the emergence of a synthesis which attempts to provide a framework to think about the role of technology in conjunction with other variables that explain why and how change happens within and outside the military. For example, Bobbitt (2002, 70–4) sought to reconcile these competing views and to demonstrate how war impacted the changing constitutional order within Europe, but, importantly, how political change also shaped the phenomenon of war.

However, achieving a balanced model in which the interaction of these variables is formed so they can be employed to explain change is perplexing. Within this literature on technology and war, William McNeill's *Pursuit of Power* (1982) stands out as a critical text on the subject of technology, war and power. It contains a central thesis that explores the role of technological innovation in Europe and its impact on politics and war. Most notably is how this feeds into a broader debate about the rise of the West. In seeking to explain why and how this happened, McNeill also looks at the example of China, which was far ahead of medieval Europe technologically, but then fell behind. McNeill considers this question and concludes that this happened because of the creation of a free market economy. In China's case, such a market system resulted in commercial and technological innovation and a host of new products and services. However, during the Ming dynasty (1368–1644), the government sought to regulate and control the market via the state. Convinced of their superiority, the Chinese not only used the state to regulate the market, they also limited trade with other countries. Within this context, technological innovation was suppressed, and all innovation was expected to flow from the centre of government (McNeill, 1982).

In contrast, the political and economic conditions within Europe allowed for the emergence of a free market system which incentivized innovation and allowed the Europeans to race ahead of the Chinese. Much has happened since this thesis was aired in the early 1980s, and the development of state-led capitalism in Japan, South Korea and, more recently, China all demonstrate that technological innovation can be crafted more directly and instrumentally by combining market and state forces. Indeed, a standard policy pursued by Western governments has been to fund defence research in the hope of creating 'spinoffs' in the form of commercially lucrative products – like the smartphone, which draws on several defence technologies, for example global positioning system (GPS), the internet, lithium–ion batteries and more recently artificial intelligence (AI).

Finally, there is a danger that following a technological trajectory exposes the study to the crime of ethnocentrism. This stems mainly from the declared purpose of much of the literature on the West and warfare, which is to explain why and how the West rose to a position of global dominance. Inevitably, much of this narrative rests on the technological advantages enjoyed by Europeans in their conquest of non-European lands (McNeill, 1982; Parker, 2016). Some historians claim this technological success has deeper roots that extend into the idea that the West possesses a distinctive culture, which allowed it to exploit technology. A strong proponent of this argument is Victor Daviss Hanson. Hanson is insistent that culture, not technology or geography, explains the success of the West as a military force. He recognizes that technological superiority is seen as an essential characteristic of the Western way of war but stresses that this does not explain the West's military success. Non-Western adversaries often had the same or better technology and were still defeated. In his view, the critical difference is that Western culture actively promotes technological innovation in a manner and on a scale that makes it unique (Davis Hanson, 2001, 9). As he explains:

> the critical point about firearms and explosives is not that they suddenly gave Western armies hegemony, but that such weapons were produced in great quantity ... in Western rather than in non-European

countries – a fact that is ultimately explained by a long-standing Western cultural stance towards rationalism, free inquiry, and the dissemination of knowledge that has its roots in classical antiquity and is not specific to any particular period of European history. (Hanson, 2001, 19)

The same arguments might also be applied to how war resulted in the creation of one of the most evolved forms of political power in the form of the nation state. In essence, why wouldn't non-Western states emulate this political format in their lands? The notion of a Western way of war has been ridiculed (Black, 2004; Lynn, 2008). Given the intellectual risk of pursuing yet another study on technology and war, some attempt must be made to explain how this is to be done so that the study does not stray unwittingly into the problems highlighted.

Revisiting the question of technological determinism

It is not just military historians who have expressed concern over an undue reliance on the use of technology to explain the history of war. Historians of science and technology have also been engaged in a fierce debate about the dangers of technological determinism. According to Walton, the concept of technological determinism has been an important theme in science and technology studies for half a century (2019, 4). For example, Ede asserts that the whole idea of technology and the assumption that it plays a positive role is loaded with implicit biases that serve to distort the truth. He describes two particular problems, which he calls 'progressivism' and 'presentism'. The first is the idea that the past existed only to produce the present. The second is the belief that the past can be judged by the standards and knowledge of the present. Progressivism is the 'belief that there is a kind of arrow of development from primitive to highly developed civilisations and societies' (Ede, 2019, 5). Underlying both these potential biases is the idea that history is nothing more than the product of a certain kind of technological determinism. The principal exponent of this idea was Karl Marx, who claimed that technology shaped a society's mode of production, which, in turn, played an instrumental role in shaping societies' social relationships and governance. This economic model assumes that medieval technology brought about feudalism and that the industrial revolution was a necessary precondition for the existence of capitalism and liberal democracy (Helibroner, 1967, 336).

There is no consensus on the question of technological determinism. Misa provides a partial explanation for this difference in views. He believes technological determinism has become more complicated because: 'historians, as well as sociologists of technology, have taken steps to achieve the difficult feat of showing technology at once as socially constructed and society shaping' (Misa, 1988, 308). He argues that this misconception stems from the level of analysis adopted by the person interrogating the role played by technology in shaping the world around us. Historians and sociologists tend to adopt a macro perspective of technology which allows them to see technology as playing a causal role in the process of historical change. In the case of the history of business and labour, a micro level of analysis means that they reject the central role played by technology in shaping history. Division also exists over whether technology is detached from other forces and completely independent of other factors (Misa, 1988, 308).

It is essential to realize that opinion on technological determination has both supporters and critics, but fundamentally they are both asking the same question: where does technology come from? Is it a consequence of society, or is it directly sourced from science – is it independent of the social domain? (Hacker, 1994) Others argue that technological development has become a supra historical phenomenon, is independent of other variables such as societal culture and, because of its increasing acceleration, it is affecting all areas of life. Reflecting on technology in the US in the 1970s, Langdon Winner (1977) believed that technology had become the primary structuring phenomenon of the modern age.

Technological determinism is a pernicious problem for those seeking to study this construct. The central question is whether it can be used and whether we can identify a point in time where we believe technological change was able to advance autonomously, which means independently of the economy and society. Zimmerman (2019) offers a possible solution here. He argues that technological determinism does sometimes happen, contrary to the views aired by several military historians. Moreover, many of those who challenge this concept tend to apply it to a context in which technology evolved very slowly and where it competes with a host of other factors. Typically, such studies have focused on the medieval or early modern period of European history. Zimmerman believes these arguments have some merit within the context of challenging the literature on military revolutions, but in the industrial world, these arguments are greatly weakened by the prevailing evidence. As he explains: 'Technological determinism is not a disease of bad historical writing but something that must be carefully applied' (2019, 45–9). Zimmerman believes that a salient characteristic of industrial warfare was the emergence of a technological system. This system integrates a range of technologies to produce a complex weapons system, but there is no one dominant technology. In the industrial world, the pace of technological change is rapid. Why then are some technologies exploited by the military more quickly than others? It is also important to note that technology does not create change, only opportunity. New weapons and technologies require an organization and a list of instructions in the form of doctrine to exploit this capability (Shimko, 2010).

Zimmerman acknowledges that technology can be shaped by social, political, and economic factors and provides the history of radar as an illustration of this point. As he explains, the technology for radar was developed by six countries in the interwar period, but only the UK incorporated radar into a national scheme of air defence. He attributes this to the profound fear in British thinking about the impact of strategic bombing on a densely populated and highly urbanized society (Zimmerman, 2019, 51). 'By focusing on systems, rather than a single technology, we can see how technologies can be decisive in war' (Zimmerman, 2019, 52).

Ellul makes the following observation about technological determination:

> This theory is often reproached for artificially isolating several factors to give one preference. However, nobody is saying there is one cause. Instead, among the countless factors operating within a society, one fact, at a given moment, appears more decisive than the rest. This factor, in turn, has numerous sources-socio-intellectual, ideological, political, etc. ... I am by no means saying technology has always, and in all societies been the determining factor ... [but] that in our Western world (and we can generalise for the past twenty years), technology is the determining factor. (Ellul quoted by Zimmerman, 2019, 53)

Ellul's principal argument is that technology is a determinant, but it is only one of a number of determinant factors that influence battles, campaigns, and wars.

So, in theory, at least, it is possible to assert that a certain kind of technological determinism has defined particular historical moments. However, this requires that we understand that other cultural, political and economic factors might have some bearing on the realization of this capability. In addition, we need to look beyond single technologies and think more in terms of a system of capabilities that embraces multiple technologies, which in defence means the weapon system. Of critical importance here is that we also reflect on the infrastructure required to support this system, and to do this successfully depends on how we conceptualize the idea of technology.

Finally, Schatzberg (2018, 139–40) claims it is possible to use the idea of technological determinism in the military domain as most technology development in this sphere has been protected from market pressures and the competitive development of similar technologies by belligerents since the late 19th century.

How we define technology will also have a strong bearing on whether or not the problem of technological determinism will surface. In essence, discussion of what constitutes technology has caused it to move beyond the hardware and now includes the institutions, organizations, management and doctrine created to employ the capability in question. As a result, it has become essential to think about the social and political contexts that also shape military technological change (Ellis, 1975; Quigley, 2013). The definitional problems associated with technology will be explored in the next chapter, but my central point is that a case can be made for a form of technological determinism that is limited in scope and time, which is the approach adopted in this study.

Addressing the problem of ethnocentrism, presentism and progressivism

While there is a consensus that war played an instrumental role in the creation of all types of states throughout history, both within and outside of Europe (Tilly, 1992, 10–14; Tin-Bor Hui, 2005; Fukuyama, 2011, 86–112), there is a consensus that too much attention has been focused on the European experience of state formation. As a result, there is a desperate need to widen the lens of enquiry within this debate so that non-European examples are brought into the discussion in the hope that a more balanced and less hubristic view might emerge. So, an important question I need to address is why I have chosen to focus on Europe and more broadly the idea of the West. My intention is not to celebrate the past military glories of the West or explain its rise via the 'barrel of a gun'. This worldview is also deeply flawed because it cannot explain why Western military power has largely failed as an instrument of war in recent campaigns; clearly, war is about more than advanced technology or the export of democracy. However, my interest in the West stems from my interest in technological innovation and the emergence of what are described as disruptive technologies specifically but not exclusively in the military realm. Technological innovation of this kind has been the West's principal comparative advantage both economically and militarily and, as such, it is simply not possible to ignore this geopolitical space.

One of the more exciting aspects of today's geopolitics is the entrance of China and India into this innovation club, but they are new entrants into military research, and India has historically been heavily reliant on the likes of Russia to provide technology transfer to build modern weapons. However, it has made a concerted effort to develop its own science and technology base, and defence has played an important role in shaping this policy area. China is further down the road in terms of its military science and technology and is now presenting a challenge to US and Western technological superiority in strategic areas such as AI but, currently, much of its arsenal is comparable to but not ahead of the United States and, according to some, it is largely based on stolen Western intellectual property.

It is also evident that Western states continue to believe that technology acts as a significant force multiplier in war and the wider economy, and are therefore eager to maintain their technological edge. The vanguard of this revolution is the US. It invests a staggering sum of money in defence, over US$800 billion in 2022. More important is how much it spends on defence research; this has increased by 24 per cent in real terms over the past ten years (Stockholm International Peace Research Institute, 2022), and its focus is on the next generation of weapons. Unsurprisingly, the US is an important focus in this study. I have also chosen to use the UK as another example from which to draw. My rationale for selecting this medium-sized power stems from the following. First, in relative terms the UK has relied heavily on technology to compensate for its lack of mass in terms of the size of its armed forces. It has therefore maintained a strong investment in defence research extending over many decades. This aspiration is clearly expressed in its latest defence policy, which stresses the importance of technological innovation to generate the capabilities required to allow it to play a meaningful role in NATO and more widely in protecting UK interests outside of Europe. Finally, like the US, the UK has a long and intimate relationship with war both in the past and the present, and provides a large and varied evidence base on which to draw.

The second reason why I have focused on the West is because an important part of this study addresses the interaction between society and technology. There is an important cultural aspect to this, which makes it considerably easier to do this with a culture that I am familiar with and understand. Susskind, in his study of technology and its impact on politics, confirms that different cultures approach the utilization of technology in different ways, which means it is important to understand how these cultural nuances have shaped a society's exploitation of technology (Susskind, 2018, 36)

In an ideal world, a concept of technological determinism also needs to be free of the notion that there is an end which equates to the creation of a modern Western society with all that this implies in terms of its own cultural superiority. The question is whether it is possible to construct this concept so that it is not laden with value judgement but can be employed to provide objective evidence of a process of change. This question is addressed more fully in the final chapter, which most definitely rejects the vision of the positive, progressive journey technology is taking us on. At its most extreme, we appear to be moving towards a future which looks less like Star Trek and a lot more like Mad Max. I exaggerate for effect here and, perhaps by 2060, developments in nanotechnology will transform the world from a place based on competition and conflict for scarce resources to a world of abundance as we acquire machines that can literally make anything we want literally from thin air. But this study represents a direct challenge to the idea that Western culture or civilization is superior to any other culture.

Revisiting the importance of technology

This book on technology and war differs from what preceded it by being up-to-date with the latest bout of technological change. Much has happened in the technological realm over the past ten years and if this change is going to be as profound as has been claimed, we need to take stock and think about the downstream effects of the current technological transformation on the nature and conduct of war and the forms of politics and political organizations it might give rise to. To recap, despite the problems highlighted regarding the use of technology as a variable in explaining change, the main argument of this book focuses on the role of technology in shaping the war–state relationship. The principal reason why I have taken this approach is because of what is currently unfolding in the world we live in. A view is emerging that we are on the precipice of a technological paradigm shift in human existence, which will affect every aspect of our lives (Harari, 2017). This revolution will apply equally to the military domain (UK Ministry of Defence, 2022). It is frequently said that we always think the latest change constitutes a revolution, but only when we look back in time are we able to make a proper and informed judgement. However, the available evidence challenges this cautious interpretation of history. Below is a list of the range of technologies which are now coming to the fore.

These technologies form the foundation of what Schwab has called the 'fourth industrial revolution' and fuse the physical, biological and digital domains (Schwab, 2016, 1–8). The revolution presents a new challenge because, in relative terms, we are witnessing a dramatic acceleration in the speed of technological change. This increase in speed is significant because accelerating technological change 'shortens the life span of ideas, business models, and market positions' (Dobbs et al, 2015, 31–2). This development is also relevant to defence and the institutions of the state. Today the speed and breadth of technological change is proceeding at such a pace that military bureaucracies and indeed other institutions are struggling to adapt, leading one observer to speak of the emergence of an exponential gap between these new technologies and the processes required to exploit and govern their use (Azhar, 2021, 9).

The rapid increase in computing power as expressed in Moore's Law is the best evidence of how technological innovation is currently accelerating. In 1965, Intel founder Gordon Moore observed that the number of transistors on an integrated circuit had been doubling every 18 months. As a result, computing power between 1965 and 2015 doubled every two years. That has now slowed to a doubling of power every 2.5 years. However, this rate of decline will be checked as new materials, for example, graphene, are created that allow more computing power to be packed into ever smaller spaces. Equally important is the rise of quantum computing. This will result in the creation of computers that are millions of times more powerful than they are today. It is also essential to recognize that it is not just the power of integrated circuits that is doubling every two years or so. According to Kurzweil (2014), this trend can also be seen in those technologies that become digital. He refers to a general law of accelerating returns that is currently in play and affects several products, including AI, robotics, quantum computers, biotechnology, material sciences, networks, sensors, 3D printers, augmented and virtual reality, and block-chain. We are witnessing the process of positive feedback, which is causing the pace of innovation to increase. In his view, 'we will not experience 100 years of progress in the 21st century – it will be more like 20,000 years of progress (at today's rate).' He believes technological advance, in general, is characterized by exponential rather than incremental change. Moore's Law, a measurement of computational power, is the latest but not the most exclusive example of this exponential growth. Kurzweil (2014) asserts that we can also see exponential growth and double exponential growth. This 'means that the growth rate is itself growing exponentially'. In essence technology creates new technologies, and as a result, the speed of technological innovation, which began slowly, increases the further forward in time we go. This means the pace of technological change today, compounded by all the innovations that preceded it, is occurring on an unprecedented scale and speed. Inevitably improvements in the life cycle of technologies reaches a limit. A case in point is the performance and cost of integrated circuits. Moore's law demonstrated that the power of microchips doubled every eighteen months. However, we will soon reach a point when it will no longer be possible to squeeze any more transistors into the physical space of integrated circuits, which will impose a severe limit on the potential power of computer chips in the near future and could mean the end of Moore's Law. However, before this happens, a paradigm shift can occur, which enables exponential growth to continue. In the case of computing power, the advent of quantum computing promises to be such a paradigm shift. A normal computer uses chunks of binary information in the form of ones and zeros to calculate and solve complex problems. This means traditional computers are only capable of processing one bit of information at a time. A quantum computer uses quantum bits.

The difference can be explained in the following terms. Think of two options for flipping a coin: heads or tails. This is how binary code works. In the case of quantum computing, the coin is constantly spinning so that both sides flash at once. This ability to calculate ones and zeros simultaneously increases the speed and power of the computer; quantum computers could be hundreds and possibly millions of times faster. As such, these machines will be able to carry out new tasks that traditional computers cannot undertake; for example, they will be able to break the most complex form of encryption and it is claimed address profound existential problems such as climate change (Marr, 2022, 24).

This is a crude summary of how quantum computing operates and how it represents a hugely more powerful system than its binary predecessor. IBM's Deep Blue computer beat Gary Kasparov at chess by examining 200 million moves a second; a quantum machine would look at a trillion potential moves before deciding on the best option (Diamandis and Kotler, 2020, 28). The development of integrated quantum circuits that power quantum computers will ensure that Moore's Law continues to hold true into the future, meaning exponential growth in computing power very quickly. As Diamandis and Kotler (2020, 30) explain:

> A fifty qubit computer as 16 petabytes of memory. That's a lot of memory. If it were an iPod, it would hold 50 million songs. But increase that by a mere thirty qubits and you get something else entirely. If all the atoms of the universe were capable of storing one bit of information, an eighty qubit computer would have more information storage than all the atoms in the universe. (Diamandis and Kotler, 2020, 30)

Based on their analysis of the state of the technology market of ideas in 2015, they identified 12 technologies they believe will challenge the status quo in the future. I have combined their 12 with technologies identified by others to create a longer list of those technologies that are expected to be potential game changers in the future (Schwab, 2016; Diamandis and Kotler, 2020; Azhar, 2021):

1. Artificial intelligence;
2. Robotics;
3. Quantum computing;
4. Next generation genomics;
5. New advanced materials;
6. New forms of energy storage;
7. Advanced techniques in oil and gas exploration and recovery;
8. Renewable energy;
9. Autonomous vehicles;
10. 3D printing;
11. Mobile internet;
12. Internet of things;
13. Cloud technology;
14. Automation of knowledge work;
15. Lasers; and
16. Virtual reality.

In sum, claims made by the military that we are on the precipice of a paradigm shift in the conduct of war, primarily precipitated by rapid technological change, are supported by an emerging consensus within the business and industrial world.

In contrast to past revolutions, this change's defining characteristic is its speed. Three forces are driving this. The first is the exponential growth of computing power, which feeds into an array of product sectors and services. Second, the rapid pace of change is compounded by the way in which these accelerating technologies are converging with each other to amplify the intensity and scale of this change. The third reflects the creation of an environmental setting that is conducive to perpetual technological innovation. One of the principal benefits of technology is that it has transformed our existence from focusing on the day-to-day challenge of how to survive to something far more comfortable. In essence, technology has enabled us to move from the bottom of Maslow's Hierarchy of Needs, which focuses on fulfilling basic physical wants, to the top of this scale of development – captured in the idea of self-actualization. As a result, we have more time to think, which is a vital precondition for the emergence of civilization.

The rate at which people adopt technology is also speeding up. It took 38 years for 50 million people to start using radio, 13 years for the same number of people to use television, four years for 50 million to get the iPod, and three years for the same number to access the internet. Facebook users climbed to over 50 million users within a year and Twitter took only nine months to reach the same number (Dobbs et al, 2015, 42). In the case of the internet, the number of users increased from 2.5 billion in 2013 to 4.66 billion in 2020. It is predicted that by 2069 every person on the planet, over 10 billion people, will have access to the internet. Innovation is also allowing us to perform even more complex tasks rapidly. As Diamandis and Kotler (2020, 67–72) point out, AI collapses discovery times from years to weeks; quantum computing compounds the speed of invention; and 3D printing shortens the development and production cycle.

According to Marr (2022), two things make the fourth industrial revolution different from previous episodes of this kind. The first is that earlier revolutions were based largely on one major technological advancement. In contrast, the current revolution is driven by several technological trends, including what he calls the 'datafication' of the world, which refers to the growth of computing in our day-to-day lives. In addition, we are now also contending with the growth of AI, the emergence of a distinct cyber domain and the dramatic expansion of the internet of things. Each of these technologies on their own would constitute a revolution in their own right, but their interaction has increased the rate of advance. For example, the enormous leap in AI has been made possible by the explosion of data, the growth of which is due to the proliferation of smart devices, which constantly generate data. This development leads to the second outstanding trait of this revolution: it potentially has no end. Innovation in one sector leads to change in another and the emergence of what Marr describes as exponential growth which carries on. 'Crucially, these tech trends, which are mingling and mixing together, will disrupt every industry, every organization (big and small), and even many aspects of society' (Marr, 2022, 22).

Azhar (2021) observes that dramatically faster timescales in innovation are not confined to the computing world. He identifies three other domains that are also following this trajectory: renewable energy, biology and manufacturing. Evidence of this transformation can be seen in the dramatic reduction in prices in these sectors, which are falling by a factor of six or more every decade. He claims that 'between these four key areas – computing, energy, biology and manufacturing – it is possible to make out the contours of a wholly new era of human society' (Azhar, 2021, 39).

A range of different sources have identified current technological change happening at a scale and speed normally associated with past industrial revolutions, which is an important signifier. Most important is the belief that these technological changes will cause transformation across the full spectrum of human activity, including the art of war.

These domains – political, economic, social and military – are not silos but feed off each other to produce synergistic effects, which are evident in such things as war and the state. In 2018, defence analyst Michael O'Hanlon observed that the speed of technological advance relevant to military innovation was likely to be faster in the next 20 years than it proved to be in the previous 20. At the front end of this fast paced revolution was the growth in computer power, AI, robotics and cybersecurity, which broadly fits the list of revolutionary technologies listed above (O'Hanlon, 2018, 5).

One possible source of information when thinking about how this technology will impact the military realm is to draw on the literature on disruptive technologies. These are described as:

A technology or collection of technologies applied to a relevant problem in a manner that radically alters the symmetry of military power between competitors. This technology immediately outdoes the policies, doctrines and organisations of all actors. (CNAS, 2014, 11)

In the military realm, the following technologies fall into this category:

1. artificial intelligence
2. lethal autonomous systems
3. hypersonic weapons
4. directed energy weapons
5. quantum computing. (Fitzgerald and Sayler, 2022)

Disruptive technologies challenge prevailing military orthodoxy regarding the best weapons to have, the optimal strategy to win a war and the duration of the conflict.

For example, does a new technology make existing weapons systems, such as tanks, obsolete? In what way do these new technologies alter the offence/defence balance? How are these capabilities best integrated into existing force structures? How are they best employed conceptually? What level of training is required to ensure maximum efficiency is achieved in the use of a new weapon? How high or low are the technical and financial barriers in acquiring these new arms? The question of how the military plan to use these technologies is explored in Chapter 5.

All of these issues also have an important bearing in helping us to understand the nature of the war–state relationship. For example, are current technological trends promoting the decentralization of political and military power or causing the opposite – the centralization of power in the military realm? This will depend on how decisive these technologies on the battlefield are, their cost, the technical complexity involved in their manufacture, and the level of training required to employ these systems. The belief that the prime determinant of the war–state relationship is economic cost masks a host of other factors that need to be addressed to understand the connection between politics and war.

In addition to the potential technology disruptors listed above, we also need to think about the prospects of human augmentation between machines and humans both physically and cognitively. However, perhaps the area of greatest interest is the connection between AI and robotics which it is feared will lead to the deployment of autonomous killing machines.

One might take the view that as these technologies already exist and, as they are now firmly part of the present, an investigation into the future should rest on the next new technological revolution. However, I think this position is problematic largely because the potential of these capabilities is yet to be realized. If we think back to the first industrial revolution, which was based on the rise of steam power, it took many decades for this technological change to permeate the military domain and it was not until the mid-19th century that we saw its effect on the conduct of war, which was some 100 years after the first industrial revolution began (see Chapter 3). Although the time between the onset of the next revolution and its impact has shortened (see Chapter 4), it is possible to see a delay between the start of a new technological change and its wider impact.

This observation also holds true in the case of the fourth industrial revolution. For example, 3D printing as a technology exists but has yet to really take off in facilitating a manufacturing revolution. Similarly, while AI exists we have yet to see the real potential of this technology take effect, which has prompted a huge debate about the viability of our existing economic model and if it has a future (see Chapter 7). The delay between the onset of the fourth industrial revolution and its impact is compounded in the case of war because of the infrequency of this event. This is important because the absence of war limits the consumption and use of technology and forces the military to resort to endless speculation about how emerging technologies will impact on future conflict (see Chapter 5). Consequently, there remains considerable scope to debate and speculate on how current technologies will evolve and change the world in which we live, especially in the military domain.

Finally, I have no doubt that the current technological foundations of the fourth industrial revolution will be eroded, and a new technological paradigm will emerge to replace it. However, as has been said the next tectonic shift is not expected to happen until sometime in the 2060s and this will be based on nanotech, which it is argued will take humanity out of a world of scarcity into one of abundance, a development which, if true, will have a profound impact on politics, economics and war. However, projecting this far out into the future presents certain risks and there is a strong correlation between the decline in the accuracy of a prediction and how far forward you look into the future. This is largely because of the lack of evidence that can be used to extrapolate from to build a credible scenario of the future. Therefore, a basic but important premise of this study is that there remains considerable potential to investigate the impact of current and emerging technologies on the political, economic and social domains, and especially in regards to war, where the effect of technological change is not always easy to discern.

Connecting war, technology and the state

There are numerous definitions of war and the etymology of the word is explored in greater detail in Chapter 4 of this book. However, it is important to acknowledge differences of opinion exist on what constitutes an act of war. Angstrom and Widen (2015) note that few states use the term war when declaring the use of force, because they do not want to be seen as the aggressors in a conflict. They also highlight that definitions of war vary as analysts focus on different aspects of what war is. These range from war as a political instrument, to a philosophical debate on war's existential nature, to more prosaic empirical measures of war as defined in terms of numbers of people killed in a conflict. All of these perspectives are useful but they can also be contested (Angstom and Widen, 2015, 13–32). However, the most important aspect of war's many faces is that this activity is embedded in violence. This raises another important question when thinking about war: what kinds of violence do we include when we talk about war? Typically, we address this question by focusing who we believe is capable of using violence in a purposeful way and why they chose to do this.

Who wages war and for what reasons has driven the philosophical debate on the question of what war is for many years. For example, do we confine our definition of war to nation states? If we accept that non-state actors can wage war, then which groups should be included? Terrorists, insurgents, warlords, drug cartels and/or gangs (van Creveld, 1991)?

A distinction between war and other forms of violence, which might help in providing clarity, is that historically war was usually, but not always, conducted on a large scale and required an extensive organization to wage it. After all, as Macmillan explains: 'it is a clash between two organised societies' (2021, 17). Macmillan goes further and argues that on a general level it may be characterized as the use of organized violence by a stable political community to achieve a political objective set by those who govern in the name of that community (2012, 18-20). This allows the definition of war to be narrowed to larger political groups including states and large scale insurgencies, but not gangs, because they are not judged to be stable political groups. Their motives are also questioned in that they fight for largely narrow economic interests that benefit only their group. As such they are not fighting for the interests of the larger community of which they are a part. In the case of what might be loosely called terrorism, we can argue that while their declared goals are intended to benefit a political community, they often lack the scale and mass to cross from the threshold of violence into the domain of war.

This broad conceptualization of war reflects the prevailing reality that formal, organized states are not the only types of political organizations that can wage war. So, for example, the UK's Ministry of Defence recognizes that states are not the only political groups that can legitimately wage war. It defined war as 'a state of armed conflict between different countries or different groups within a country' (UK Ministry of Defence, 2022, 1).

Viewing war in this way also makes it easier to focus on the historical connection between war and state formation and the interplay of technology. The importance of this triad has varied across history, but it is now an opportune moment to explore this connection because current technological change is challenging our traditional conception of war and the state. There are three reasons why this is happening. The first is that technology is increasingly replacing humanity's central place in war. The second is that technology is creating new ways of waging war that pose an existential threat to the whole notion of what constitutes a war. The third is that technological advance is also leading to a decentralization of power with technologies that can be used to wage war becoming more accessible.

In looking at the problem of technology substituting humanity in war, we can see this happening at a number of levels. The most obvious manifestation of this trend can be perceived in the increasing automation of the battlefield. If robots replace humans, then this implies a massive change in our understanding the nature and conduct of war. The same problem arises off the battlefield in a military headquarters. Here the concern is how to exploit the advantages of AI to increase the speed of decision-making without making the slower human commander and his team redundant (Keegan, 2014; Coker, 2014; Kissinger, 2018). Of particular importance here is the focus on war's nature. Typically, we acknowledge that technology will change war's conduct, but its nature remains fixed. For most military historians, the importance attached to war's nature stems from our reliance on Clausewitz's conception of the nature of war and his focus on an analysis of the interaction between human cognition and the brutal reality of battle and the broader stresses imposed by war on the morale of the commander and the army (von Clausewitz, 1976, 577-616). However, if you take the human out of war, this concept ceases to have any relevance, and war becomes very different. Are we moving towards such a moment? This question forms an essential line of inquiry in this book.

Another level concerns the conduct of war and the increasing shift away from battle in the conduct of war. Recent Russian and Chinese activity, particularly the exploitation of new technologies, is challenging this paradigm of war and causing us to rethink what constitutes an act of war. Most important here is how media and social media are being used as forms of political warfare, which increases the overlaps between war and peace and poses an interesting question about where we place the boundary between peace and war now. The question posed is whether this represents a form of warfare. If it does, how will these other instruments be combined with more traditional tools of war? If an enemy can be defeated through deception and disinformation before its forces are engaged in battle, what are the implications for the war–state relationship? Will technology, traditionally the West's best friend, be turned against it, causing political disruption and potential fragmentation? In essence, will war in the future result in the weakening of the state? If so, what conditions are needed to enable such an attack to reach a tipping point to break the cohesion of this political entity?

The democratization of technology and its increasing availability to everyone also reduce the barriers to entry into war. Perhaps the best demonstration of this phenomenon are computer hackers who, via a laptop, can cause huge disruption to large-scale organizations including states. A good example of this was the ransomware attack 'Wannacry', which caused huge problems for the delivery of public health in the UK in 2017. If war can now be waged by even small groups what does this mean for the war–state relationship?

In sum, my central aim is to explore the future connection between war and the state and explain why and how current technological change is impacting the war–state relationship and, most importantly, how this interaction will unfold further into the future. Given the presumption that war made the state, have we reached a point when the Western obsession with technology as a means to achieve profit and success in war now poses a threat to the logic of the war–state relationship? The central purpose of this book is to explain why and how Western states might be acutely vulnerable to this next evolution in the war–state relationship. It is important to establish a clearer understanding of the key terms employed in this book to achieve this goal. In this case, I am specifically interested in the concept of the state. This clarification is necessary because it will allow us to see the many connections between the technology, war–state relationship.

Tilly has observed that the term state is frequently applied to various political systems, including city states, empires and even theocracies (1992, 2). He defines states in general as a distinct organization that controls the principal concentrated means of coercion within a well-defined territory and, in some respects, exercises priority over all other organizations operating within the same territory. However, the principal interest of this study is in what he defines as national states. This is different from its predecessors because 'a national state or modern state then extends the territory to multiple continuous regions, and maintains a relatively centralized, differentiated, and autonomous structure of its own' (1992, 131). This conforms closely with Max Weber's definition of the state, which 'is a human community that claims the monopoly of the legitimate use of physical force within a given territory' (Gerth and Mills, 1946, 77). However, a state is more than a hollow space in which an armed Leviathan watches over a passive population. A state is also the sum total of what Fukuyama calls political order. This is largely determined by three categories of institution: the state, the rule of law and mechanisms of accountability (Fukuyama, 2011, 23). According to Tilly, the state is defined as a 'hierarchical, centralized organization that holds a monopoly on the legitimate use of force over a defined territory' (Tilly, 1992, 23). The rule of law is defined as a 'set of rules of behaviour reflecting a broad consensus within the society, binding on even the most powerful political actors' (Tilly, 1992, 24). 'Accountability means that the government is responsive to the interest of the whole society – what Aristotle called the common good – rather than to just its narrow self-interest' (Tilly, 1992, 24). This last applies mainly to democracies. The institutions of the state concentrate power and allow it to govern, enabling and legitimizing governance and accountability to constrain the state's power. Liberal democracies are believed to be unique in that all three institutions exist side by side and, most importantly, have reinforced the legitimacy of the power exercised over those governed.

What the state does and how it functions is a direct consequence of how that order operates and how it decides to fulfil basic goals such as generating stability and safety for the population who live under the armed might of the state. This means thinking about the state as more than a mere bureaucratic structure. It is also more than just the enforcer of the law. We often conflate the government and the people with the notion of the state, and while this might be a consequence of fuzzy thinking, it does highlight the reality that we cannot separate one from the other. As such, when thinking about the state, it is also important to recognize the role the state plays in creating and sustaining the legitimacy of government via the diverse range of services provided by it to the people and that it is closely entwined in a political and policy-making process in which its function is to deliver the promises made by those elected to rule. Even when not elected, populist authoritarian leaders will still rely on the state bureaucracy to play the game of 'bread and circuses' which aims to buy the people to win their hearts and minds.

Consequently, the modern state is a vital part of the political legitimizing process that all governments, democratic or otherwise, seek to promote. This last point is crucial because our understanding of the state extends into society. Perhaps this is best illustrated through Clausewitz's concept of the trinity in war, which depicts politics within the state as a dialogue between the government, the institutions of the state in the form of the armed forces, and the people. The resulting interaction between them shapes the political object of the war, how it shall be fought and how it will end. However, it also provides a convenient framework to demonstrate the impact of current technological change (von Clausewitz, 1976).

The structure of this book

The following chapters in this book build on these questions and/or themes. Chapter 2 addresses the question of technological determinism in war and politics and seeks to calibrate the relationship between technology, war and the state by situating it within the context of probably one of the most discussed periods of military change, the early modern period within Europe. The principal argument in the second chapter confirms the position taken in the introduction that technology is one of several variables that precipitated a change in war and the process of state formation, but that more credible explanations that challenge the role of technology are discussed and these pose important questions about the utility of technology as an agent of change.

In Chapter 3 I contend that the importance of technology increased dramatically in the conduct of war from the 19th century onwards for three reasons. Schumpeter's economic analysis of capitalism and its relationship to technology demonstrates that four long economic cycles in the industrial revolution led to ground-breaking changes in the mode of production in little more than 100 years (Schumpeter, 1994). These changes took time to permeate the consciousness of the military, which is not surprising given the innate conservatism of armed forces. But in the second half of the 19th century, we witnessed a greater enthusiasm to employ new ideas and technologies in war. This chapter seeks to explain why there was such a change in the mindset of the military and how this impacted the connection between war and the state until the end of the Second World War. Mass industrialized war in the 20th century emphasized quantity more than quality and required the mobilization of society and the economy via the state. The demands of war also resulted in the state expanding into the provision of education and healthcare to ensure the population was fit to wage war. Even liberal Britain succumbed to this view of the state. These features eventually became the defining characteristics of what Hables Gray called 'modern war' (1998, 128–47).

Chapter 4 focuses on the advent of the nuclear age and the changes it precipitated in the organization and conduct of war. Most important here is what impact the nuclear revolution had on the war–state relationship. The existing literature perceives a significant decline in this connection as the incidence of war reduced in frequency in the Western world. This chapter challenges this view in several ways. First, it revisits our understanding of war and what constituted an act of war against the broader background of a possible Armageddon. Most important here is how technology fits into this new strategic paradigm. Precisely, how did the military use it in this more constrained setting, and what were the implications for the war–state relationship? This is a particularly interesting moment in history because of the hugely significant role played by military, scientific research, which for a short time became the vanguard of the technological revolution that swept through society starting in the 1980s.

Chapters 5, 6 and 7 are linked by the adoption of a Clausewitzian framework of analysis which Colin Gray recommended in any study of future war. As he explains, 'Future warfare can be approached in the light of the vital distinction drawn by Clausewitz, between war's 'grammar' and its policy 'logic'. Both avenues must be travelled here. Future warfare, viewed as grammar, requires us to probe probable and possible developments in military science, regarding how war could be waged. From the perspective of policy logic, we need to explore motivations to fight war' (2005, 39).

Chapter 5 looks at the grammar or science of war that has emerged in Western military doctrine in response to current and perceived geopolitical and technological threats the West is likely to confront over the next decade or so. Essentially, this is their manual on how to fight the next war and win. Within it are numerous assumptions about the nature of international politics, who the next enemy will be, how they will seek to negate the military effectiveness of the West's military system and how this can be countered. The important question here is how accurately this vision of future war addresses the emerging reality we confront.

Chapter 6 also focuses on the future grammar or science of war and explores how far the Russian invasion of Ukraine validates the assumptions underlying Western doctrine about the future war. It explores how technology has changed war and if those changes conform with expectations mapped out in Western military doctrine.

Chapter 7 then sets out the future policy logic of war. Its principal focus again is on the technological domain, and it seeks to explore how the next wave of technology will pose a challenge to the Western military's vision of war's grammar in the future and a challenge to the internal coherency of the state. The central argument here poses a question regarding the future role of technological innovation. In the recent past, it is possible to assert that the process of invention and the implications of these new practices played an important role in state formation in the Western world, but have we reached a point where current and future changes in the technological domain are creating a new policy logic in the realm of war while at the same time challenging the cohesion and functioning of the state?

2

Technological Determinism and Debates about State Formation in Early Modern Europe

This chapter explores the origins of technology and the war–state relationship. Although I recognize that each of these domains has a long history, the starting point of this chapter focuses on the transition from the late medieval world into the early modern period of European history and then extends into the late 18th century. The selection of this timeframe is dictated, in part, by the existing literature on this subject, which sees this as a definitive period in the relationship between these three domains. Thus, there is a consensus that war played an instrumental role in the creation of all forms of states throughout history, both within and outside of Europe (Tilly, 1992, 10–14; Tin-Bor Hui, 2005; Fukuyama, 2011, 86–112). However, this resulted in the creation of a range of types of states. As Tilly observed, the term state is frequently applied to various polities, including city states, empires and even theocracies. Thus, he defines states in general as 'a distinct organization that controls the principal concentrated means of coercion within a well-defined territory, and in some respects exercises priority over all other organizations operating within the same territory' (1992, 2). However, as I have explained, the principal interest of this study is in what he defines as national states. This is different from its predecessors because 'a national state or modern state then extends the territory in question to multiple continuous regions, and maintains a relatively centralized, differentiated, and autonomous structure of its own' (1992, 131). This conforms closely with Max Weber's definition of the state: 'is a human community that claims the monopoly of the legitimate use of physical force within a given territory' (Gerth and Mills, 1946, 77).

In the case of the modern state, the connection between war and the state was complicated by the intervention of a new variable in the form of technology. In broad terms, it is claimed that technological and organizational innovation in early modern European warfare, both on land and at sea, precipitated a military revolution which, in turn, created a political revolution represented by the emergence of the modern state (McNeill, 1982; Downing, 1992; Roberts, 1995; Knox and Murray, 2001). This has been the dominant narrative in discussions about early modern war and state formation in Europe. As a result, the debate rarely questioned the notion that there had been a revolution and instead settled on tactical issues such as precisely when the military revolution started and whether it is possible to talk in terms of a revolution when discussing a process that took a century or more to unfold (Black, 1991; Rogers, 1993; Parker, 2016). However, this technological focus has been challenged by a small cabal of military historians who emphasize the importance of political and social changes in shaping the evolution of war (Black, 1994; Knox and Murray, 2001; Parrot, 2012). In the case of Knox and Murray, they see the rise of the state as a consequence of wider systemic forces that led to greater centralization of power and this in turn changed the character of war, precipitating a series of smaller revolutions (2001, 6–12). Bobbit sought to reconcile these competing views and to demonstrate how war impacted the changing constitutional order within Europe, but also, importantly, how political change also shaped the phenomenon of war (2002, 70–4). Importantly, social scientists have, in general, been content to accept this broad interpretation of history (Tilly, 1992; Toffler and Toffler, 1993; Mann, 1986; Hirst, 2001) and their theories of war and the state rested on the logic of innovation in the military realm leading to political change. This chapter draws on these three perspectives to challenge the prevailing view of the war–state dynamic in the early modern period. Of particular importance is the role assigned to technology in precipitating a military revolution that led to a political revolution manifesting in the emergence of the state doctrine of absolutism in the 18th century. I aim to provide a better understanding of how significant or not the role of technology was in shaping the war–state relationship in this period. This requires that we acknowledge the significance of politics, culture and economics as equally important drivers of change and that technology is merely one variable that explains change in human history. In sum, the principal purpose of this chapter is to demonstrate the limitations of technology as a cause of change.

In thinking about this subject, one of the first and most important things we need to do is define what we mean by the term technology. Nevertheless, almost immediately, we are confronted by a problem. Precisely what kind of activities are included or excluded from this construct? Some definitions are pretty broad, focusing on how things are made or done, but nearly every form of human activity could be squeezed into such a definition, making it almost meaningless; others are perceived to be too narrow and miss important material and non-material processes that contributed to technological change (Shatzberg, 2018, 1).

Is it even appropriate to use a term that, according to Schatzberg, only came into general use in the second half of the 20th century and apply it retrospectively to early modern Europe (2018, 214–17) As he explains, the concept of technology was utterly absent from early modern discourse (2018, 11). The danger is then that we impose motivations and a form of logic alien to those who lived in the times being scrutinized.

The Society of the History of Technology (SHOT), which is an interdisciplinary academic association that specializes in the study of the impact of technology on society, also takes issue with standard definitions of technology because they reinforce the sense that technological change happens without reference to other social and political forces, and presumes innovation is inevitable and follows a particular trajectory of development. In SHOT's view, technology is not autonomous, but the result of choices made 'very often in disputes over power manifested in registers of politics, gender, race and inequality'.[1] In the past, this problem has manifested itself in one of three ways, all of which reinforce the view that history follows a predetermined course, which illustrates the superiority of Western technology and culture. Ede (2019) describes this phenomenon as 'progressivism' and 'presentism'. The first is the idea that the past existed only to produce the present space we now occupy and that the past can be judged by the standards and knowledge of the present. Progressivism is the 'belief that there is a kind of arrow of development from primitive to highly developed civilizations and societies' (Ede, 2019, 5). Underlying these potential biases is the idea that history is nothing more than the product of a certain kind of technological determinism. The principal exponent of this idea was Karl Marx, who claimed that technology shaped a society's mode of production, which, in turn, played an instrumental role in shaping that society's class structure and social relationships. This presumes that medieval technology brought about feudalism and that the industrial revolution was a necessary precondition for the existence of capitalism (Helibroner, 1967, 336). The big question, of course, is how accurate this reductionist view of human history is. Given these problems, definitional debates about technology have led to the adoption of less prescriptive definitions. Thus, SHOT defines it as 'the sum of all methods by which a social group provides itself with the material objects of their civilization'.[2] Schatzberg (2018, 1) defines it as 'the set of practices humans use to transform the material world, which involves creating and using material things'.

---

[1]   www.historyoftechnology.org

[2]   www.historyoftechnology.org

The problems highlighted above have been particularly pronounced in the study of military history and are one reason its academic credibility has suffered. The most ardent critic of the dominant role given to technology in military history is Jeremy Black. He observes: 'many problems are faced when discussing technology and its role in warfare, not least that we have been educated since the 1960s to expect technological solutions to be definitive' (Black, 2004, 104). It is not just military historians, however, who are guilty of assigning too important a role to technology when explaining the evolution of war; scientific historians have also played their part. Perhaps the most famous was an observation made by Lyn White, who claimed the introduction of the stirrup into Europe in the 7th century CE led to the creation of feudalism.

His argument is that this happened because of the introduction of what was then a new technology into medieval Europe in the form of the stirrup. Attached to the saddle via leather straps, these metal hoops hung on each side of the horse and rider placed his feet in them to control the movement and direction of the horse. The stirrup was invented in China and then introduced into Europe via the Avars and the Byzantines. In battle its principal function was to ensure the rider stayed on the horse. The ability to capitalize on the speed and mass of the horse allowed cavalry to charge in a disciplined group, typically with the lance, which made them formidable on the battlefield. The addition of heavy armour, both for the rider and the horse, meant that cavalry forces were mobile, heavily protected and could deliver the kind of physical and psychological shock sufficient to break infantry armies and or cavalry forces that did not possess the stirrup. As a result, the use of the stirrup spread quickly across the European continent and battles came to be dominated by heavily armoured cavalry, which were called knights.

How then did this technology lead to the rise of feudalism? It appears the cost of acquiring and maintaining this capability was extraordinarily high. It reinforced the dominance of a warrior class sustained by a predominantly peasant population and a social and economic system that consolidated the knight's dominance politically and militarily (White, 1962). While there is general agreement that the stirrup made the knight a more potent weapon in war, many question its role in the creation of a social and economic order. It has been explained that, in some regions of Europe, the emergence of feudalism was not accompanied by the rise of the knight (Roland, 1993). Others have observed that the dominance of cavalry in this period needs to be seen and judged in relation to the poor quality of infantry rather than the excellence of cavalry, a phenomenon that was due to poor social cohesion that stemmed from political rather than technolgical causes (Murillo, 1999, 57–8).

The early modern period is also presented as an example that contains key elements that allowed historians and social scientists to claim a connection between technological change, war and state formation. In scrutinizing these claims, it is first important to provide a baseline, setting out the principal characteristics of politics and war in the period preceding the onset of changes in both domains and then identify the causes. In political terms, medieval Europe was characterized by a diffusion of power. Various types of states in the form of feudal lordships, emperors, kings, and city states co-existed within a defined geopolitical space. Cutting across all of these vertical forms of political power was the Roman Catholic Church, whose wealth, access to necessary forms of knowledge, both material and non-material, and ideological legitimacy accorded to it as God's representative on Earth combined to create a complex set of political and social interactions (Spruyt, 1994, 3).

Politics also played an instrumental role in shaping the character and conduct of war. The collapse of the Roman Empire in the 5th century CE left small and fragmented communities to fend for themselves, and this led to the creation of new social structures that combined elements of the Roman world with the political and military culture of the groups who occupied the former Western part of the Roman Empire. According to Ede (2019), an example of this fusion was the persistence of the concept of comitatus in the post-Roman world. Such a system imposed a military obligation on those awarded lands to govern on behalf of their lord and, in return, they answered a call to arms and went to war if necessary (Ede, 2019, 113). In the past, these nobles would probably have fought as infantry in a Warband led by their king or lord, much as Harold and his army of Huscarls and the Fyrd did at the Battle of Hastings in 1066. However, the introduction of the stirrup and the subsequent rise of cavalry meant that warfare in the medieval period became largely dominated by armoured cavalry in the form of knights, supported by levies of peasant infantry (Bean, 1973, 205). Because of the huge expense of arming and training these knights; it is estimated that a knight's armour and weapons equated to ten years of pay for an archer (Rogers, 1993, 245). In addition, the huge logistical challenges of supplying an armed horde as it moved across the country also limited the size of armies and the distances they could cover (Mann, 1986). Movement and control of space were further complicated by castles, which allowed the local lord to dominate his lands and, on occasion, provided safe refuge when the nobility challenged the monarch's rule (Bean, 1973, 207). These factors combined to make war a ponderous affair dominated by sieges and in which battle was rare. Moreover, even when armies clashed on the battlefield, knights, eager for ransom, fought more often to capture other nobles than destroy the opponent's army in a Clausewitzian sense (Rogers, 1993, 246). Most importantly, the character of warfare in this period sustained a political map that, in the 15th century, consisted of over 5,000 distinct political units (Macmillan, 2018). Tilly (1992) notes that the desire to centralize political power, impose control and create fewer and larger political units was a pronounced tendency in the process of European state formation. What facilitated the realization of this long-nurtured dream was, he believes, caused by the introduction of gunpowder into European warfare (Tilly, 1992, 70–5). The idea that gunpowder changed the complexion of war is a view that none would contest. For example, Francis Bacon declared that the modern world was created from three inventions: the printing press, the compass and gunpowder. Thus, the 18th-century historian, Edward Gibbon, claimed that gunpowder 'effected a new revolution in the art of war and the history of mankind' (Black, 2013, 71). Given the number of subsequent innovations triggered by this chemical revolution, it is hardly surprising that its introduction into Europe sometime in the 13th century was perceived as the start of a paradigm shift in the conduct of war. More recently, Black (1994, 4) has argued that the stirrup and gunpowder represented important military revolutions in the character and conduct of war. How then did this precipitate the creation of the modern state? As has been said, a range of theories and ideas have been put forward, linking changes in the conduct of war and a particular kind of state formation to the emergence of new technologies. Given the salience placed on the idea of technology as a cause of change, it is important to map out this process of causation. Below is a broad synthesis of the literature on military revolutions, which all suffer from the problem of technological determinism. To redress this imbalance, the remaining part of the chapter explores the importance of non-technological factors in explaining the rise of the modern state in the early modern period. This will help by providing a wider context to judge the overall importance of technology as a facilitator of change in the war–state relationship.

Revolutions within revolutions

Debates about a military revolution in the early modern period all agree the initial spark that precipitated a dramatic change in the conduct of war was caused by the introduction of gunpowder into Europe, a development that van Creveld heralds as the start of the age of machine warfare (1989, 82). Black's history of this innocuous substance locates its discovery as being in China, sometime in the 9th century CE. From there, knowledge of how to manufacture it was discovered in India and possibly via the Mongols it first entered Eastern Europe in the 13th century (Black, 2013, 70–2). Another suggested point of entry into Europe was via the Middle East and the last Crusades (Ede, 2019). The development of this technology and the means to employ it, in terms of cannon and firearms, evolved slowly and in an ad hoc fashion but, as Black explains, once gunpowder emerged in the West 'a different mindset approached the questions of improvement, which was eventually pursued with considerable energy' (Black, 2013, 72). The first references to handguns appeared in texts as early as the 1370s, and by the end of the 14th century, the first cannons were being used. It is important to note that both the Chinese and Arabs also developed guns at about this time, but there is general agreement that it was indeed within Europe that this technology was adapted most effectively to the demands of waging war (McNeill, 1982, 81). Improvements in the combustibility of gunpowder, its system of ignition and metallurgical improvements resulted in the creation of more robust firearms and cannons capable of firing larger projectiles over a greater distance. This evolution occurred slowly due in part to cultural resistance to how this technology changed warfare (van Creveld, 1989, 86–7) and the time it took to demonstrate it was a more effective military technology than the weapons it was intended to replace – the bow and muscle-powered siege engines (Black, 2013, 72–83).

The defining moment when the gunpowder revolution seemed to change war is symbolized by the French invasion of Italy in 1494. According to Howard, Charles VIII's army represented the first modern army because it was composed of three elements: infantry, cavalry and artillery (Howard, 2009, 38–40,). Of these elements, the artillery arm presented the most dramatic change in the conduct of war and, in so doing, challenged the survival of feudalism. This happened because improvements made in the design and construction of siege guns in Europe. According to Rogers (1993, 266), the critical developments happened between 1400 and 1430. During this time, guns evolved from being large pots resting on a wooden frame, which fired large metal bolts into something that more closely resembled what we today would recognize as artillery. Innovation in construction resulted in the creation of longer and stronger barrels capable of firing a larger shot. According to van Creveld, by the late 15th century, such cannons could fire a round stone a metre in diameter and weighed almost a metric tonne (1989, 87). The effectiveness of these siege guns was demonstrated by the fall of Constantinople to the Turks in 1453. The principal limitation of this weapon – known as the bombard were its size and weight. These early forms of cannon were forged using techniques borrowed from the moulding of large church bells. As such, bombards were so heavy it was sometimes easier to cast the metal for the gun at the location of the siege. The stress caused by the detonation of firing a large stone meant their rate of fire was limited to one or two shots per day. The utility of this form of siege artillery was greatly enhanced because of an arms race between the kingdoms of France and Burgundy between 1465 and 1477. Gun founders in both countries discovered that much-improved casting techniques resulted in smaller guns that could fire iron rather than stone cannonballs. Most importantly, these could inflict the same damage as bombards, which were three times the size. As a result, siege guns became mobile and could be moved relatively easily across country on specially designed gun carriages. It is claimed the design for these guns remained unchanged from 1477 until 1840 (McNeill, 1982, 88). Although bombards had already changed the balance between attack and defence, the creation of mobile siege artillery undermined established power relationships, and it is claimed to have led to sweeping societal changes. The effect was to produce a centralization of political power. This happened first within states as only the monarchy could afford to construct siege artillery, which then allowed them to destroy the local nobility's basis of power – the castle – and it is claimed to have played an instrumental role in facilitating the rise of absolutism in 18th-century Europe (Rogers, 1993, 273) and ultimately the creation of the modern state (Bean, 1973, 203). The international consequences of this momentous change were demonstrated first in northern France, where English controlled towns in northern France rapidly fell to the new artillery employed by the French. Whereas it took Henry V ten years to conquer Normandy after Agincourt in 1415, in just over a year, the French reduced 60 castles and recaptured the province. Even more spectacular was Charles VIII's invasion of Italy in 1494. The Italian city states could not resist and quickly came to terms with the French drive to seize the throne of Naples. The eventual siege of Naples demonstrated the power of the new artillery. A city that withstood a seven-year siege fell to the French after only eight hours of bombardment, reducing its walls to rubble (Duffy, 1979, 8–9). As McNeill observed, 'the major effect of the new weaponry was to dwarf the Italian city states and to reduce other small sovereignties to triviality' (1982, 89). However, Italian engineers had been experimenting with a range of measures to counter siege artillery's effectiveness since the 1440s. So, for example, in 1500, the Pisans, besieged by Florentine forces using bombards, built earthworks to reinforce their walls. This second line of defence presented a significant obstacle because it absorbed the kinetic energy of the cannonballs and remained largely intact. Earthworks, ditches, ramparts and bastions came to form the technological answer to the question of how to stymie the effectiveness of siege artillery (Duffy, 1979, 15). These were then constructed to provide the defender with the maximum opportunity to exploit its own firepower by creating clear zones of fire around the fortifications and ensuring all parts of the defence could provide supporting fire so that if an enemy attacked one wall, it would be subject to withering fire from adjacent walls. This is why forts assumed a star-shaped pattern. This development then forms the second part of the technologically deterministic picture of the evolution of war in the early modern period. New forms of fortification, termed the *trace italienne*, emerged to counter the artillery revolution. The fortification revolution came to dominate warfare in the early modern period. However, in contrast to castles, which facilitated the decentralization of political power, the enormous cost of building this elaborate system of defences, including the artillery and human resources it needed, ensured only the wealthiest states could afford this military architecture, and this reinforced the power dominance of the state and reinforced its ability to secure a public monopoly on the use of force (Bean, 1973, 207; McNeill, 1982, 90). Fortifications spread across the geopolitical map and dominated European warfare from 1525 to 1790. Bean (1973, 204) asserted that the size of states was partly a function of economics in that economies of scale resulted in reduced costs for a public good like defence and, with the fortification revolution, those costs had to be amortized over a larger population. Paradoxically, the power of fortifications also served to check the growth of empire in Europe, as demonstrated by the efforts of the Hapsburgs to create a single imperial entity stretching from Portugal to Austria. In this case, the new fortifications simply made conquest so much more difficult. Equally important, this technology checked the expansion of the Ottomans both in Malta and along the border between the Holy Roman Empire and the Ottomans in Hungary (McNeill, 1982, 91).

Why then did the artillery revolution lead to a profound change in the structure and philosophy of European governments? To answer this question, we need to turn to Geoffrey Parker's thesis on the military revolution in early modern Europe. Parker (2016) sees a direct correlation between the combined effects of the gunpowder revolution and the fortification revolution and the rise of absolutism within the European state system. The key variable in explaining this connection between the military and the political was the increase in the size of armies starting in the mid–16th century (Parker, 2016, 45–6). Why did this happen? In the case of the fortification revolution, his argument is straightforward. The *trace itallienne* and the fortifications that followed were located at strategically important points, and this meant an invading army could not ignore them as it advanced into a country for fear that the enemy garrison would cut its supply line if it chose to ignore it. However, defending or besieging a fortification of this kind was incredibly labour intensive. Concerning the defence, the fort required a large garrison of several thousand soldiers to defend it effectively from attack; such garrisons might account for over half the state's army in static positions. From the attacker's perspective, a direct assault was too risky and likely to end in heavy loss, so attacking forces typically relied on a slow and deliberate siege to break the enemy and capture the city. However, to do this, the attacking force first needed to erect two sets of walls. The first faced the defenders in the city and were erected out of the range of the defender's artillery. The second circumvallation faced away from the fortified town and was constructed to prevent the enemy from breaking the siege. To illustrate the scale of this, a town with walls of 1500 metres in length might require an attacker to build and guard a double circumvallation that was 40,000 metres in circumference (Parker, 2016, 13–14). The most pronounced effect of this development was the decline in the incidence of battle, which briefly surged during the artillery revolution as it became clear that rulers could no longer rely on castles to defend their lands. The fortification revolution restored the superiority of the defence and, as a result, wars were dominated by sieges, not battles. Parker (2016, 16) cites Marlborough's campaigns, which consisted of over 30 sieges but only four battles. Ribot-Garci (2001, 46) notes that the overall effect was to make war less decisive, which, in turn, caused conflicts to become more attritional and protracted. These changes made war increasingly expensive as rulers sought to sustain larger forces in wars that dragged on.

The second source of change taps into a debate first articulated by Michael Roberts' thesis on the military revolution in the early modern period and concerns the firepower revolution on the battlefield. Roberts' (1995) principal argument was that changes in the battlefield tactics between 1550 and 1660 precipitated profound changes in politics, society and the economy. Parker accepted this idea but sought to demonstrate that the revolution started earlier than 1560 and was not confined to just the Netherlands and Sweden but included those territories controlled by the Hapsburgs.

It was generally recognized that early firearms were not as effective as bows and longbows. They had a limited range and a low rate of fire and performed poorly in wet weather. An archer could fire ten arrows in a minute and be accurate at a range of 200 metres. In contrast, an arquebus had a range of only 100 metres and took several minutes to load (it took 40 actions to load and fire an arquebus). Parker (2016, 17) believes bows and crossbows were quickly replaced by firearms because the latter had greater penetrating power than bows, and it took a fraction of the time to train a soldier to use an arquebus. Black (2013) is sceptical that early muskets or handguns had greater power than a steel crossbow, but there does seem to be a consensus on the relative ease of training a person to use a handgun or arquebus compared, say, to the longbow, which took many years of training. Black speculates that this technology replaced existing firepower forms because they cost less than steel crossbows in the 15th century. He also sees a connection between its use in the defence of fortifications discussed above and its acceptance more widely in the military domain (Black, 2013, 82).

According to Rogers, 'the military revolution which fills the century between 1560 and 1660 was, in essence, the result of just one more attempt to solve the perennial problem of tactics – the problem of how to combine missile weapons with close action; how to unite hitting power, mobility, and defensive strength' (Roberts 1995, 13). The introduction of large numbers of soldiers with firearms required them to combine with other types of weapons to create a combined formation less vulnerable to cavalry and infantry armed with edged weapons. The answer to this question was to create mixed formations consisting of musket-armed infantry and soldiers armed with a traditional weapon, the pike. These usually operated in large formations, but the tactical challenge was to ensure these units could transition from firing to forming a literal hedgehog of pikes if forced to fight hand-to-hand. The Dutch under Maurice of Nassau first devised a tactical system that relied on close order drills to facilitate the rapid movement and deployment of pike and musket-armed infantry in a manner that allowed them to maximize the use of firepower by deploying in line rather than dense columns. Such tactical innovation ensured they could fire in volleys and reform into a dense pike phalanx when required. Alternatively, by employing a technique described as countermarch, such mixed formations could also be used offensively in battle as troops were taught to move through the ranks of a formation to advance, fire, retire, reload and advance. This created a prolonged and continuous barrage of musket fire. To achieve this goal, it was necessary to break down the size of units from 3,000 soldiers to smaller sections of 30 and ensure each section had its own commander so the unit could respond to a change in orders. The Dutch, because they were only rarely exposed to battle, failed to realize the potential of this new system, and so it was the Swedes, under Gustavus Adolphus, who took this tactical doctrine and refined it to allow them to pursue the dream of a decisive battle in the Thirty Years War (Parker, 2016, 20). According to Bobbitt (2002, 99), this tactical and organizational innovation, conceived by the Dutch and implemented by the Swedes was as crucial to the transformation of the state as the fortress revolution.

To summarize, the logic of the military revolution school runs as follows. The introduction of gunpowder into Europe in the late 13th century resulted in the artillery revolution in the late 15th century. This was then followed by the fortification revolution, which happened at about the same time. Finally, tactical and organizational reform resulted in the firepower revolution on the battlefield of early modern Europe. Each of these technical and organizational changes in the infrastructure of war supposedly caused a change in governance captured in the rise of the modern state.

Mann shows how military innovation in the early modern period resulted in radical political change. The artillery revolution allowed the monarchy to break the domestic nobility and disarm them. At the same time, the huge cost of artillery ensured that only states could acquire this capability, reinforcing its stranglehold on the monopoly of violence. These trends were reinforced by the subsequent fortification revolution, the cost of which confirmed the state's control over the means of violence. Finally, as mapped out by Roberts (1995), the firepower revolution reinforced the shift from an individualistic form of warfare based on personal skill and a culture that emphasized the warrior's ethos to one which relied on mass and tightly controlled organization to achieve success on the battlefield. The skills required to wage war could no longer rely on the privately trained nobility and or segments of the wider population drawn into the business of war as and when obliged. Again, this created significant barriers to entry into the military market for all organizations and groups except the state. According to Mann (1986) , the demands of war signalled the end of the feudal state. Reliance on levies of largely untrained peasants was obsolete. Similarly, city states in areas such as Italy also struggled to compete. Simply put, they could not find the money needed to maintain their independence in siege warfare. Finally, the drive to improve the effectiveness of firepower on the battlefield also centralized political power. The demands imposed by complex drills and formations made it impossible to rely on militias and facilitated the rise of mercenaries who had time to refine and hone their fighting skills. The combined effects of these changes, both technical and organizational, expanded war both physically and temporally as these changes reinforced the power of the defence and, as a result, warfare became protracted and frequently indecisive. This, then, produced positive feedback as states had to generate and field forces for extended periods beyond the normal campaign season, and this also reaffirmed the need for a permanent professional military under the control of the state. The outcome of these changes in the declared relationship between technology, war and the state was the creation of an expensive military, which required the creation of a bureaucratic infrastructure capable of extracting the manpower and resources, in the form of conscription and taxation to sustain the military machine. The size of this task can be measured by looking at the growth in the size of armies and navies. Using France as an example, during the 1480s, Louis XI maintained a force of 45,000 soldiers. In the 1550s, Charles V had 150,000 men under his command. At the end of the 1600s, Louis XIV had an army of over 400,000 men (Ribot-Garcia, 2001, 38). A similar trend was apparent in the maritime domain. England's naval tonnage increased from 30,000–40,000 tonnes in 1558 to 196,000 tonnes in 1700 (Parrott, 2012, 12). It was not just that armies and navies increased in size during the early modern period. As important was the impact of inflation on state budgets. According to the Spanish Minister in the Netherlands in 1596:

> If comparison be made between the present cost to his Majesty (Phillip II) of the troops who served his armies and navies, and the cost of those of his fathers the Emperor Charles V, it will be found that, for an equal number of men, three times as much money is necessary today as used to be spent then. (Parker, 2016, 61)

In essence, the demands of war emphasized the advantages of larger centralized administrations, which is why European states moved towards the creation of a state system dominated by highly centralized forms of administration and political control (Mann, 1986, 452–5; McNeill, 1982, 117–84). In other words, technological change in the conduct of war played an instrumental role in creating the Absolutist state in 18th-century Europe.

The existing literature presumes a link between the rise of absolutism and the rise in the cost of defence. As Delbruck (1990, 223–4) explains:

> The entire socio-political situation of Europe was transformed with the new military organization. The standing army was the point of contention in the struggle between princes and their estates of the realm, the factor that raised kings to absolute rulers on the whole continent. ... As a prerequisite, or perhaps we should say a side effect of the great change in the army, there developed a new administration of the state, a bureaucracy whose mission it was to collect the taxes required to maintain the army and, by careful handling of the economic conditions and finally the entire welfare and agriculture, to make the country as productive as possible.

Bobbitt (2002) also sees a strong connection between military revolutions and state formation in early modern Europe and links several of these to defining moments when states experienced rapid political change. He does this by accepting competing claims made by military historians that a military revolution happened at a particular moment as being valid and avoiding the arguments between them. For example, he sees a close connection to Parker's thesis, which focuses on the fortification revolution and state formation in the 15th century. He also acknowledges the connection between Roberts' military revolution and the rise of the modern state in the 17th century and explains why this form of governance prevailed over other political forms that existed at this time, for example, Italian city states and most notably, the Holy Roman Empire. As he observes: 'The strategic innovations of ever more expensive fortress design and complex infantry fire crushed those constitutional forms that could not adapt to exploit those innovations' (Bobbitt, 2002, 101). So, for example, the Hapsburg dynasty's efforts to create an imperial state were doomed. 'Such an entity simply could not manage sufficient control of its domestic resources to maintain standing armies capable of the prolonged campaigns required to vindicate dynastic claims that were often geographically remote and politically fraught' (Bobbitt, 2002, 105).

Antithesis

What then is wrong with this thesis? This technologically deterministic picture of state formation in early modern Europe can be challenged on the following grounds.

A point stressed by Black (2004), is that the Europeans did not have a monopoly on gunpowder, firearms, artillery or siege guns, and these capabilities were employed by various non-European states. However, this did not always result in the creation of highly centralized systems of governance whose principal purpose was to create and sustain a war machine – the modern state. Black explains there were numerous polities not suited to the creation and maintenance of large permanent military forces because of climate, terrain or low population density, which highlights the point that the mere presence of gunpowder was never going to be sufficient on its own to produce a military revolution. Moreover, in some states, creating a large military establishment was not a political priority (Black, 2004, 71–2). Consequently, state formation assumed a very different path in some states, even though this technology was present. A good illustration of this 'deviancy' from the presumed norm is China, which invented gunpowder and possessed crude firearms and cannon at about the same time as its European counterparts. However, its relationship with technology and its militarization was complicated by a range of cultural and political factors, in particular the importance of Confucianism within government and its mistrust of what we today would describe as the operation of the free market and the rise of a private entrepreneurial class. Such activity and the individuals conducting business in this way were increasingly regulated by the government, suppressing innovation within the Chinese economy. At the same time, there was deep mistrust between Chinese civil servants and the military, and the former remained determined to keep the military on a tight leash. This was achieved in part by limiting the capabilities the military could procure so that they could not conduct offensive military operations. So, for example, during the Sung dynasty, which observed the invention of gunpowder, the government did make a concerted effort to encourage the invention of firearms. This happened because of the declining technological difference between the Chinese and 'barbarians' caused by the occupation of northern China by these forces, and a need arose to maintain and increase this technological gap. However, at this stage, firearms and crude forms of artillery fitted in well with a defensive strategy, and so the government encouraged their development. Finally, although the Sung dynasty supported the development of gunpowder, heavy reliance was also placed on non-military activities like bribing the frontier tribes to stay away (McNeill, 1982, 27–39).

This takes us back to a point made at the start of this chapter: the connection between war and state formation. How acute the external threat is perceived to be to the state also plays a role in the extent to which a polity might seek to exploit military technology. Thus, Bobbitt (2002) argues that profound changes in the war–state relationship are a consequence of epochal wars. These are wars:

that produce fundamental challenges to the state. A warring state that is unable to prevail within the then dominant strategic (military) and constitutional practices will innovate. In such wars, successful innovations – either strategic or constitutional – by a single state are copied by other, competing states. This state mimicry sweeps through the society of states and results in a sudden shift in the constitutional orders and strategic paradigms of an epochal war. By this means, a new dominant constitutional order emerges with new bases of legitimacy. (Bobbitt, 2002, 67)

Epochal wars can be distinguished from other wars because they embrace several separate wars, punctuated by moments of peace into one long war. Examples include the Hundred Years' War and the Thirty Years' War. The First and Second World Wars and the small wars folded into the Cold War are also seen as one epochal war (Bobbit, 2002, 19–21). These are existential conflicts in which states draw on whatever means is available to ensure their survival, a prominent feature of the European states system that emerged in the 17th century. Thus, Tilly (1992) notes that from 1400–1800 a significant conflict started every two to three years. The frequency of such conflagrations increased to a significant war every two years between 1800 and 1944. The intensity of these wars, as measured in battle deaths also rose from just under 3,000 per year to during the 16th century to more than 223,000 deaths per year in the 20th. What is most interesting is how many of these wars from 1600 involved European powers. Since European states formed the core of the world's great powers from 1600–1900, it is safe to assume they formed the lion's share of these wars in this period (Tilly, 1992, 67).

State formation is seen as a result of how best to extract the resources needed to survive and conquer. Hendrix Spruyt (1994) emphasizes the relative insignificance of technology as the principal determiner in state formation. He observes that politics went through what he describes as a tectonic shift at the end of the Middle Ages as the cross-cutting jurisdictions of feudal lords, emperors, kings and the Catholic Church gave way to the sovereign state governed by a single authority. This evolution changed the structure of the international system by basing authority on the idea of the sovereign state (1994, 3). Interestingly Spruyt recognizes that the modern state was one of several possible constitutional orders that emerged toward the end of the Middle Ages, and all were possible contenders for creating a new international order. The underlying driver that he believes caused these political forms to emerge was the role of trade, not war. In his view, trade led to various coalitions between monarchs, the nobility and towns in different countries, which led to new political forms to improve trade. Three specific political forms were identified: the city state, the city league and the territorial state. In the competition between these new forms, the territorial state proved the most efficient in generating wealth. This was facilitated by the elegant simplicity of having a single source of political power in the form of a monarch and proved superior when compared to the diffuse collective decision-making of the city state or league (1994, 7).

Two tangential but still essential questions posed are to what extent the processes of state formation outlined were peculiar to Europe and, second, if a military revolution or revolutions of the early modern period caused the creation of the modern state, then how do we explain the different political complexions of these states and, vitally, the variation in the concentration of political power within them?

According to Bobbitt (2002), a vital factor that supported the creation of the modern state system within Europe, which he argues began in Renaissance Italy, was the role played by the Catholic Church. This provided an overarching institution that facilitated the development of a comprehensive set of legal practices, norms and customs that rulers accepted because they legitimized their claim to govern. It also provided a means through which rulers with competing claims could arbitrate through the good offices of the papacy if required. The existence of the legitimating power of the Church helped bound a region characterized by considerable local variation in terms of custom and culture. Most important here was the regulation and control of political power as represented by the electoral process determining who became Emperor of the Holy Roman Empire. The Church also provided a readymade bureaucracy which monarchs frequently drew on to govern. Paradoxically, the imperatives of war resulted in creating a political order that also challenged the Church's legitimacy in the long term as competence in governance and statecraft rather than bloodline became increasingly important (2002, 76–84). Downing (1992) also believes that the military revolution, as articulated by Roberts, depended on how it impacted existing forms of governance within late medieval European society, which emphasized the importance of local forms of political representation – an embryonic form of democracy found only in Europe. He describes this as 'medieval constitutionalism'. This form of governance stretched from England to Iceland and Scandinavia in the north, Spain and Italy in the south, the Low Countries, and most of what was the Holy Roman Empire (Downing, 1992, 4–5).

Downing's thesis also attempts to address the second question concerning the types of regime that emerged in Europe. In his view, the key variable that explained the progression towards democracy or the drift towards authoritarian rule depended not on the economic system and the politics that gave birth to it but on how states responded to the challenge of military modernization, which he believes began in the early modern period (1992, 10). Downing is keen to stress that:

> the perspective advanced here is not a narrow technological determinism that argues that military modernization and modern warfare lead simply to a more form of state and social organization. The key to the rise of military-bureaucratic absolutism is not modernization and warfare themselves but the mobilization of resources to fund them. (Downing, 1992, 14)

Funding the creation and sustainment of this military operation created a political crisis within states, and the responses of rulers in terms of funding the rising cost of war, in turn, shaped the future political trajectory of that state.

This connection between the rise of a new military system and political form has been challenged. Black (1994), for example, recognizes the importance of the expanding bureaucratic apparatus in states like France and Prussia but believes this drive for efficiency was not so much a consequence of technological change. In this case, it stemmed from the desire to increase the size of national armies and navies, which were largely political choices dictated by monarchs' eagerness to pursue hegemony or prevent it. Thus, innovation in war was largely driven by social and political developments (Black, 1994, 10). Of importance here was the impact of the Treaty of Westphalia, which all but ended the control exerted by the Holy Roman Empire over more than 300 German principalities. Equally important, it legitimized the right of rulers to determine the internal governance within their states, including the state's official religion. In doing so, it set a safety mechanism intended to stop hugely destructive religious conflicts like the Thirty Years' War from occurring again. This period of relative stability that followed this disastrous war resulted in greater prosperity, facilitating the state's expansion and military capabilities. It is within this more peaceful setting, starting from 1660 and ending in 1720 that Black believes the most critical military revolution, as measured in the increase in the size and effectiveness of European forces, took place (Black, 1994, 9–10).

Parrott (2012) also challenges the thesis of war leading to the creation of the modern state and again points to broader systemic forces that facilitated the creation of this political form. He highlights the importance of religious reformation and the drive by rulers to impose norms and beliefs on their populations that resulted in the creation of distinctive national cultures. He also considers the role played by economics in precipitating a shift in state power. Of particular importance here was the combined effect of rapid population growth across Europe, the impact of inflation on living standards, increasing concentrations of wealth and hence rising inequality and the rising social and political tensions this caused. He notes that an essential response to these trends was for ruling elites to lean towards more authoritarian forms of governance. Finally, he observes the power of those who ran state bureaucracies and the personal and organizational self-interest that pushed them to expand their bureaucratic domains, which indirectly led to the expansion of the state (Parrot, 2012, 12).

Two things then become apparent. First, it is believed by some that state formation was a consequence of both local conditions specific to Europe at this time, including the developments experienced in the conduct of warfare. However, the role assigned to technological innovation in war and its role in the process of state formation is contested. Further evidence of the perceived weaknesses of logic that finds a technological change in war leading to political change can be seen from examining the presumed connection between the military revolution and absolutism. As has been said, the problem here lies in the fact that the military revolution resulted in the creation of modern states that are profoundly different. The first and most straightforward way of challenging this thesis is to question the presumption that absolutism was ever the political reality claimed. For example, Black (1994) argues that past historians have exaggerated states' military and administrative capabilities in the 17th and 18th centuries. He believes the expanded armed forces during this period developed in a manner that did not challenge society's social and political reality (Black, 1994, 235).

Parrot (2012) also contests the claims made about the rise of absolutism as a direct consequence of the demands of war. As he explains, if the state is defined in terms of possessing a monopoly on the use of force within a territory few if any states before 1792 satisfied this basic measure of state power. Mercenaries and military contractors remained an important element of warfare throughout the early modern period (Parrott, 2012, 308).

The importance of non technological factors in war and state formation can be seen from a cursory glance from two examples from medieval Europe. First, Rogers argues that an essential evolution in the conduct of war and one which proved to be a defining characteristic of modern war and the state was the rise in the importance of infantry in the 14th and 15th centuries as measured in the defeats infantry-based armies inflicted on knights towards the end of the medieval period. These military successes were not achieved as a result of new technologies but rather the resurrection of an older generation of weapons in the form of the longbow and the pike. The effectiveness of infantry in relation to cavalry was caused, in part, by their involvement in prolonged conflicts such as the Hundred Years' Wars which resulted in increased military professionalism. He believes that the rise of infantry armies exerted an essential social and political impact on European society and that the enfranchisement of commoners and their increasing political power directly resulted from their increased involvement and effectiveness in warfare (Rogers, 1993, 252). This coincides with Howard's (2009) view, who also sees the rise of infantry armies during this period as an important development but believes this happened as much because of the increasing social cohesion of lower stratas of medieval society as their increased involvement in war.

Second, evolution of warfare in 13th and 14th century Italy reinforces the importance of the political as opposed to the technological in explaining state formation and confirms how prolonged exposure to war produces significant macro-political changes. In the case of Italy, the conflict between city states, regions, the papacy and other external powers created an endless succession of wars that precipitated a profound change in the civil–military relationship within these policies. Traditionally, the defence of cities and towns relied on the use of the local militia, but the constant pressure of war demanded the presence of a full rather than part-time military. According to Covini (2001), this imposed a significant economic cost on those who served and resulted in the more prosperous strata of city states offering money instead of military service. She also notes that militia troops were not particularly effective in military terms. Their equipment was often of poor quality and lacked the combat experience offered by the mercenary market. In addition, from the perspective of those who governed, mercenaries were perceived to be politically neutral in so far as they were not part of any faction involved in the internecine conflicts that erupted within and between political communities within the region.

As a result, the mercenary market in Italy grew exponentially over the next century, and most ordinary Italian governments paid for this public good via increased taxation and loans. Indeed, such was the scale of this activity that, in response to the market's demands, a new form of mercenary enterprise emerged in the 14th century – free companies. These were essentially private armies under the command of what could loosely be described in today's parlance as a warlord. Although effective, free companies were expensive to maintain and beyond the means of most city states, especially when used in prolonged conflicts. Failure to pay or the termination of a contract resulted in free companies taking matters into their own hands and plundering the property of the host population. The free companies became too powerful within the politics of Italian city states. They held a local monopoly on the use of force and consequently were able to intimidate governments with the threat of violence and disorder into awarding more extended contracts. In response, states sometimes fought wars simply to keep these mercenary bands employed. In the longer term, states gained control over mercenaries by offering longer-term contracts and through a process of absorption. Mercenaries were encouraged to put down roots in states and awarded lands and estates in return for military service. The Treaty of Lodi, ending the wars between Milan, Florence and Naples in 1454, also reduced the mercenaries' power and forced them to accept the rules imposed on them. Bureaucratic structures were created to organize and administer these armed groups, and city states like Milan and Venice became skilled at managing this privatized form of violence (Covini, 2001, 12–23).

It is also important to note the significance of constant competition between states within Italy caused these entities to maintain and expand their permanent military formations, and this needs to be taken into account when looking at war and the formation of the modern state. Covini notes that in the case of city states like Florence and Venice, which were more interested in commercial rather than territorial expansion, they too were forced by the threat posed by other city states like Milan, which were focused on territorial expansion, to secure control over the wider region to protect their cities from attack. Thus, in her view:

> The link between war and state formation is particularly evident in Italy, where territorial aggregation also took place at the regional level. The political formation that took place at the end of the fourteenth century shows clear elements of statehood, for example, territorial expansion, the tendency to centralize various functions, the growth of administrative, fiscal, diplomatic, and military bureaucracies, and the greater powers of coercion on the part of governments. (Covini, 2001, 27)

The same elements were at work in terms of how war shaped state formation in early modern Europe. Thus, Garcia sees the creation of the modern army and state between the 15th and 18th centuries as symptoms of deep and intensive competition between Europe's monarchies. Of equal importance were their efforts to centralize political power via the creation and expansion of bureaucracy and secure a monopoly over war and diplomacy (Ribot-Garcia, 2001, 38).

Conclusion

There is little doubt that the emergence of the state is a direct response to the threat of violence facing a political community (Fukuyama, 2011, 537). As such, war has had a central role in state formation. However, not all states are the same and the social, economic and political context within which a state exists will also determine the overall complexion of the state in terms of how centralized or decentralized power is distributed within it. This, in turn, will determine the extent to which it can mobilize for war. The last point will also reflect the geopolitical location of the state in question and how acute the threat to its security and survival is. Within this broader tapestry, I have attempted to explore the role played by technology, employed as a means to improve military effectiveness in battle on land and sea, and its role in facilitating the emergence of a certain kind of political order. War and its evolution within early modern Europe reveals a simple but elegant argument which claims that changes in military technology, largely in the introduction of gunpowder, but extending to more mundane items such as the introduction of flintlock muskets and even socket bayonets led to a variety of changes in the form and operation of military power. The principal consequence of this technological change seems to have been to increase the complexity of war, leading to a greater emphasis on professionalism across all ranks, the shift from militia to permanent military formations, and the requirement to build and sustain a military infrastructure in the form of naval yards, fortifications, supply depots, barracks, all of which precipitated an increase in the cost of preparing and waging of war. This increasing fiscal burden was accepted as a necessity and led to the growth of a bureaucracy capable of extracting the revenues required to sustain this force, which also required an increase in the authority and coercive power of the state. In addition, sustaining and running a large military also required the expansion of the bureaucracy to ensure basic administrative tasks were completed. The illustration of this model is interesting because of the cogency of the argument and the debate it has sparked in terms of the challenge directed at the argument in defence of technological determinism. This is also a useful perspective in that it shows the complexity of how and why change happens and the limitations of this argument that rests solely on technological innovation. Most important here is what followed this supposed military revolution in the form of the French revolution and the Napoleonic Wars. These wars also constituted a military revolution and ranked as one of Bobbitt's epochal wars. However, the sources of this tectonic struggle stemmed from political and ideological factors based on the ideas of democracy and nationalism, which spawned the drift towards mass armies, evidenced by the increase of armed forces on all sides, not technology. The next chapter explains why the industrial revolution resulted in the rise in the importance of technology and a form of technological determinism which steadily grew in prominence.

3

The Industrial Revolution and the Rise of Modern War

In the previous chapter, I set out the principal limitations of technological determinism to explain the evolution of warfare in Europe from the early modern period onwards. This chapter aims to explain why technology became a critical variable within the war–state relationship and how it impacted the character of war. In addressing this question, it is also essential to explain why the military became increasingly enamoured with the allure of technology and how they sought to harness it. In constructing this audit, it is helpful to think of technology's direct and indirect effects and why and how it shaped the war–state relationship. Similarly, it is also essential to map out how these changes altered the character of conflict. The rise of modern war is significant because it allowed states to wage war on an unprecedented scale and duration, exemplified in the two world wars of the 20th century. To this end, this chapter is divided into three discrete sections. The first section addresses the question of why technology became fashionable in the military realm. The second looks at the drive for technology in war and how this affected the war–state relationship. The final section concentrates on how this increased focus on the technical means of waging war determined the character of modern war as explored via the impact of technology on strategy, operations and tactics.

I hope to demonstrate that technological innovation's importance had a profound impact on the war–state relationship during the period under scrutiny. However, its rise was facilitated by various factors, including the intensity of competition within Europe's regional security complex in the last third of the 19th century. In addition, internal organizational change within European armies in terms of their order of battle, modus operandi, and command and control, which preceded this revolution, also facilitated the incorporation of new technological enablers. Finally, I believe the principle of being able to mobilize the entire male population in a society to wage war was firmly established in the largely preindustrial era. The Napoleonic

Wars demonstrate this point as first France, and then other nations rallied the population for war.

Consequently, this action cannot be credited to the industrial revolution, even if you extend the origins of this episode back to the 16th century as has been claimed (Toffler and Toffler, 1993). As such, while I can see that the rise of modern war, especially in the 20th century, led to a profound change in the nature of the social contract between the citizen and the state, I do not connect this to technological change in the military domain. You need only look at the Confederacy during the American Civil War, a largely agrarian society imbued with a culture that was anti-technological and pre-modern, but which mobilized to the fullest extent possible to preserve the institution of slavery (Gallagher, 1977). I see this as a consequence of a political revolution, which complemented the change in the material conditions of war made by the industrial revolution, which facilitated it. At the same time, it is crucial to recognize Fukyama's (2011) observation that the modern state's evolution was shaped by the industrial revolution and the social mobilization it gave rise to. As such, we need to be aware of this broader context within which war was situated. As he explains, economic growth generated new social groups, which organized and united to form interest groups eager to impose their demands via the political process for a fair share of the distribution of wealth within society. The industrial revolution resulted in a sudden and dramatic increase in the growth rate of per capita output in those societies that experienced it. As a result, 'far more change would occur in the succeeding two centuries than in the preceding two millennia' (Fukuyama, 2011, 29). It is essential to understand this broader context and the fact that war, the state as a rational bureaucracy, and society were all changed by this revolution. These changes worked to the benefit of war-making. In particular, it created the wealth used by the state to improve education, health and social care. While these are seen as a product of an enlightened society, Tilly (1985) makes the point that they also served the interests of the state. By making provision in these domains, it ensured it could fully employ its population during times of war (Tilly, 1985, 169–86). In the case of warfighting, its consequences were more evident. However, one profound change which emerged was the prominent position taken by the state and the military in becoming the vanguard of the technological revolution that emerged towards the end of the Second World War.

In the military realm, the impact of technology can be seen by how quickly the pace of development picked up. This point is illustrated by a cursory glance at naval technology, which shows that ship design and armaments in Europe remained essentially unchanged from 1560 to 1850 (Huntington, 2005, 361). However, in less than 100 years, we witnessed dramatic change. For example, the heavily armoured Dreadnought of 1905, weighing 20,000 imperial tons, would have been unrecognizable to a sailor from the

Napoleonic Wars in a vessel made mainly of oak and weighing only 2,000 imperial tons (Hirst, 2001, 25). The speed of change in this domain of war was also unprecedented, as demonstrated by the fate of HMS *Warrior*. Built for the Royal Navy in 1861, it was the first armour-plated iron-hulled ship but was obsolete in less than 20 years as a new class of mastless warship, which allowed for mounting turreted large calibre guns, came into service in 1873. In this period, airpower emerged from being an auxiliary activity in the form of balloons to become a decisive arm in the prosecution of war on the land and at sea. In essence, the fruits of the industrial revolution permeated the military domain in the second half of the 19th century and warfare endured a succession of technological convulsions, which, for the military, meant that 'war instead of repeating itself, became an exercise in managing the future' (van Creveld, 2009, 205).

Why military luddites became enamoured with technology

McNeill observed that the 'ritual routine of army and navy life as developed across the centuries discouraged innovation of any kind' (1982, 224). However, when the industrial revolution began to impact the conduct of war, this mindset experienced a profound change – reflecting on the importance of technology and war in 1944, military theorist, J.F.C. Fuller asserted, 'weapons, if only the right ones can be found, form ninety-nine percent of victory' (1998, 186). Although a contestable proposition, this comment does reveal a sea change in the mind of the modern military. Two questions arise here. First, why did this reactionary organization come to change its attitude towards innovation and second why did this focus so heavily on technology?

The technical fruits of the industrial revolution emerged only slowly in the consciousness of Western militaries starting in the 1840s (McNeill, 1982, 224). Navies were the first to embrace the need for change with France implementing radical reform in the 1840s in an attempt to offset the UK's dominance of the seas via technological innovation. In contrast, the Royal Navy was wary of technological innovation because it did not want to have to face the task of replacing its vast fleet of wooden sail ships because some new technology threatened to make the fleet obsolete. On land the catalyst for change was Prussia. Its ambitious foreign policy but weak economic and geopolitical position led it to exploit technology to compensate for its inferior position on the political map of Europe in the 1840s. In essence, the Prussian state was outnumbered and surrounded by countries opposed to its policy to create a unified German state. Its strategic vulnerability led it to employ all the means available to overcome these disadvantages. The first of these measures rested on creating an organizational and institutional apparatus to train and mobilize its male population to supplement its small but well-trained professional army during war. Of particular importance here was the brain of the army in the form of the general staff, which promoted a level of efficiency that ensured the army was a well-oiled machine. It also sought to use technology as a force multiplier and was one of the first states to develop a breach loading rifle for general use by all its troops. This technological change, although modest in scale, revolutionized battle by allowing soldiers to reload while lying down rather than standing up, which significantly improved the chances of survival on the battlefield and meant they could sustain a higher rate of fire compared to those using rifled muskets. Given Prussia's poor geopolitical position, extensive use was also made of the railways to allow it to use interior lines of operation to mobilize and deploy against possible future enemies. Command and control of these armies were exercised via the telegraph, which allowed the high command to direct the movement of armies and corps across long distance in hours rather than days. The advantages of these innovations were demonstrated in Prussia's stunning victory against the Austrian empire in 1866 and then France in 1870–71. These military successes changed existing conceptions of power in international politics. Historically the concept was measured loosely in terms of wealth, population size and access to resources such as land or control of the sea. The more a state had of these, the more powerful the kingdom or state was. As a result, there tended to be a strong correlation between power and the size of territory under the control of a state.

Posen added a second dimension to the concept of power, which focuses on the state's internal mobilization of resources to create the capability to alter external power balances (Posen, 1984, 62). This last point is significant when looking at the relationship between industrialization and the evolution of modern war. In addition, individual weapons came to be seen to exert an influence in determining relative power balances between states. This was particularly true in naval warfare, where small technical advantages might prove decisive. The industrial revolution fundamentally changed our understanding of power in two ways. First, the wealth generated by this phenomenon challenged the distribution of power within the international system facilitating the rise of states not endowed with a large pool of natural resources. It also allowed power transitions to happen more rapidly. Organski (1968) captured this trend in his theory of power transition, which addressed a weakness in Realism that failed to explain states' rise and decline within the industrialized world. Most importantly, industrialization dramatically improved the ability of states to mobilize the resources within its defined territory to wage war.

The unification of Germany in 1871 intensified the competition between Europe's major powers. The political change to Europe's map saw the creation of a German state that was so powerful that it took a combination of France, Russia, the British Empire and the United States to defeat it during the First World War. In essence, the creation of Germany effectively led to the breakdown of a meaningful balance of power, which resulted in intense competition to restore this balance via alliances and arms racing. Arms racing in this period assumed two forms. The first and most apparent was the manifestation of increasingly large armed forces and the capacity to rapidly mobilize a vast proportion of the male population of European states in times of war. So, for example, the German army increased dramatically from 470,000 in 1871 to 748,000 by 1914. Its total strength on mobilization for war was over two million men. It was hoped that with this massive force, Germany could precipitate the rapid defeat of the French army, which stood at 1,800,000 when fully mobilized and then that of Russia, with a fully mobilized army of 3,400,000 men. Within this context, the incorporation of mass production techniques proved to be an essential foundation for the prosecution of modern war. So, for example, until the Crimean War (1850–53), small arms manufacture in the UK was primarily a cottage industry which relied on skilled artisans to manufacture pistols and rifles. By the 1850s, this system of production seemed both antiquated and highly inefficient when compared to what was called the American system of production. This system relied on machine tools to manufacture the components that made weapons. As such, it was faster and far more precise; for example, it was possible to swap the parts of one weapon with another, which was very difficult to do with weapons that were made by hand. The slowness of gun manufacture in Europe was demonstrated in the case of Prussia's efforts to manufacture the Dreyse breach loading rifle. On average, some 22,000 rifles were manufactured per year. As a result, it took over 26 years before they had enough rifles to arm their soldiers just as they entered the Austro-Prussian War. In contrast, the adoption of the American production system at Enfield arsenal in Britain in 1863 resulted in an annual production run of over 100,000 rifles. This method spread rapidly, and in 1866 in France and 1869 in Prussia, both states could reequip their armies with new rifles in just four years (McNeill, 1982, 236).

The second element of arms racing in this period was a pronounced drive to improve and invent new weapons with better performance. Although there was a lag between the start of the industrial revolution and its impact on the conduct of war, its effect was at first modest in the sense that what constituted potentially revolutionary change amounted to little more than drawing on best practice in the wider economy. Before 1880, the invention process among defence contractors and government arms factories tended to be a random process that relied heavily on the genius of individuals who were either wealthy industrialists or sponsored by such private wealth. Within this context, research and development were mainly borne by a cabal of entrepreneurs who sought to exploit the defence market. The willingness of the entrepreneur to accept financial and technical risk in a market shaped by the conservatism of the military ensured that research was conducted on a modest scale (McNeill, 1982, 278).

However, the relationship between defence and civil innovation changed at a certain point, and defence began to assume a leading role in innovation. This, then, led to greater emphasis on the performance of weaponry as much as how many of these weapons were available to wage war. The drive for performance stemmed from two sources. The first was the reliance on a military doctrine which harkened back to the Napoleonic Wars. This creed emphasized the importance of decisive battle, and all activity was geared towards achieving this goal, even though realizing this objective against a multimillion-man army was virtually impossible (Naveh, 1997). The second driver reflected a profound fear that no major military power had the economic resource to fight a prolonged war. Consequently, a rapid and decisive outcome was sought, which placed great emphasis on offensive action and the exploitation of technology to provide an advantage in war. In essence, non-military factors reinforced the compulsion to fight an offensive war and win quickly and decisively (von Bernhadi, 2015).

The Anglo-German naval arms race is the defining moment in the industrialization of war when competing states sought to develop weapons that represented a step-change in capability and made the enemy's forces obsolete. In this case, the emerging threat posed to Britain's security by Germany's insistence on embarking on an ambitious naval armament programme caused the Admiralty to put increasing pressure on naval shipbuilders to construct ships with heavier armament, and more powerful engines to improve speed, which was particularly important in allowing the development of the destroyer, a new type of ship whose purpose was to rely on speed and to engage and neutralize another technological threat posed by submarine-launched torpedoes. Heavier armour also extended to the introduction of gun-mounted turrets to protect them from enemy fire. An essential benefit of gun-mounted turrets was their ability to traverse and fire across a wide arc, as significant was the increase in the range of naval guns because of improvements in the power of propellants, which allowed bigger shells to be fired. This also caused the length of barrels to become longer. This, in turn, made it necessary to introduce breech loading rather than muzzle loading guns. The increased range of guns then required improved fire control systems that allowed ships to hit targets over 13,000 yards away accurately, a feat made more difficult as both the target and the ship firing on it were both moving at speed. These developments happened over a relatively short time, no more than 20 years and represented a significant leap in capability, which could not be satisfied by drawing on the existing pool of technical knowledge within commercial shipbuilding. As important, the financial and technical risks associated with such programmes meant that privately owned shipbuilders were reluctant to fund these developments out of their financial reserves because the cost of failure could lead to bankruptcy. As a result, the state's financial power began to play an increasingly prominent role in creating bespoke defence technologies. As McNeill explains:

> Financial problems became especially acute because of the unpredictability of costs. This, in turn, arose from the rapidity with which new devices and procedures were introduced. Over and over again, a promising new idea proved less expensive than it first appeared would be the case; yet to halt in midstream or refuse to try something new until its feasibility had been thoroughly tested meant handing over technical leadership to someone else's navy. (McNeill, 1982, 278)

McNeill called this growing encroachment of the state within the realm of defence innovation 'command technology'. This term captured a trend which was becoming increasingly apparent in the late 19th century. It presumed that states could no longer rely on the free market to generate technological innovation; this was too slow, too erratic and subject to great uncertainty. Command technology offered a solution and assumed that if the wherewithal of the state and the private commercial defence contractor's skills could be combined, innovation need not simply be subject to the whim of luck or the right person creating a new idea. In essence, command technology was intended to force innovation to happen as and when required, making this a largely state-controlled process (McNeill, 1982, 269–306).

Command technology remained confined mainly to naval procurement until the onset of the First World War. However, the drift into a total war represented the realization of the military and politicians' worst fears. The strain it imposed on all states created massive pressure for innovation, especially in land warfare, where the challenge of trench deadlock frustrated the generals' ambitions to rekindle the hope of a great and decisive battle. As a result, a concerted effort was made to apply the practices employed in naval acquisition over the preceding three decades to this problem to create a more deliberate form of invention (McNeill, 1982, 331). Existing weapons from small arms, in particular the development of light machine guns, rifle grenades and mortars, all improved the effectiveness of infantry. Airpower also emerged as a distinct domain in war as technical improvements resulted in more robust and powerful aircraft, which could carry bombs to be dropped on the enemy behind their front line. This was also a war in which new weapons were employed, such as flamethrowers and chemical warfare. The most significant of these inventions was undoubtedly the tank. This weapon was explicitly designed to deal with the topographical challenges posed by attempting to cross complex terrain dominated by barbed wire, concrete bunkers and huge shell holes, hence its rhomboid shape. This tracked vehicle became one element in the unlocking trench deadlock, as demonstrated by the Allies' use of this weapon in the war's final offensives in the summer and autumn of 1918. Its initial design and manufacture were placed under the jurisdiction of the Royal Navy. Indeed, it was also operated by naval personnel when first deployed in battle on the Somme in 1916; they were even called land-ships to emphasize their nautical sponsor. In the case of airpower, the process of command technology was also evident. As Maurice Pearton (1982) observed, evidence of this process was apparent in the case of the British aircraft industry even before the First World War, and the government took a keen interest in the development of aviation. It established the National Physical Laboratory in 1909 to carry out research on aeronautics. It also sponsored competitions to determine the optimum airframe design and best aero engines at its research facility at Farnborough. The need to push the technological limitations on airpower resulted in the state playing a leading role in the research, development and production of aircraft of all types during the war and after (Pearton, 1982, 172).

Europe's interwar period was marked by an effort to identify, capture and institutionalize the tactical lessons learned from four years of total war. During this time, new technologies from the First World War in the form of the tank, aircraft and aircraft carrier matured in technological, if not doctrinal and organizational terms. Even supposedly new forms of war, such as paratroopers, had at least been imagined in the latter stages of the First World War. As a result, military innovation in the interwar years continued. In broad terms, this assumed three forms.

The first maintained the status quo and accepted that the next war would look like the First World War. It was believed this would be a protracted and attritional conflict. Within this context, it was assumed that the first two years of war would be primarily defensive while the national economy mobilized. Once complete, a series of grinding offensives would be implemented until the enemy was defeated. This was largely the view taken by the French, but also embraced by the British in the years leading up to the breakout of the Second World War (Strachan, 2005, 216-39). This vision did not preclude technological innovation, and the British and the French invested in refining what had proved to be war-winning weapons in the form of artillery, the tank and the aeroplane. France also invested heavily in a complex system of fortifications that stretched along the Franco-German border. Called the Maginot Line after the defence minister who sanctioned its creation, this system of fortifications represented state of the art in terms of what was technologically feasible in creating a modern defence. A whole army was able to take refuge in this system. It contained sleeping quarters, dining facilities and entertainment in the form of cinemas. In terms of its defence, its principal firepower rested on heavy artillery protected in casemates and machine guns placed within concrete pill boxes. The system was connected by an internal railway, and the air pressure inside was slightly higher within the fortification to counter the use of chemical weapons. Tanks and airpower had a role to play in this defensive system. However, a great deal of reliance was placed on the impact of firepower to stymie any offensive and exhaust the enemy before launching a counter-offensive. As such, although the French produced tanks such as the Char B and Somua, which were superior in some respects to their German counterparts, no real thought was given to their employment beyond supporting a defensive strategy. In the case of the British, they experimented with armour, and two important theorists, Liddell Hart and J.F.C. Fuller, set out competing theories of how to use armour offensively (Bond, 1977; Reid, 1987). However, a cash-strapped army struggled to find answers to its questions about how best to employ new technologies. Both cases demonstrate that the presence of technology is not a guarantee that it will be employed successfully, and a range of other factors must be considered (Kier, 2017).

New technologies also emerged in the interwar period which complicated the task of understanding how to fight the next war. The most significant of these new weapons lay in the realm of the electromagnetic spectrum, for example, radar, which had a profound impact on the conduct of air and naval warfare and wireless radio, which was sufficiently reduced in size to allow command and control to be extended downward to the smallest military units in the air, sea and land domains.

In the case of radar, both the Germans and British researched this area of technology to understand how to use radio waves to identify the location of aircraft. The British made good use of this technology to create an integrated air defence system, which proved vital in the Battle of Britain as the Royal Air Force and Luftwaffe fought to achieve air superiority. Using the 21 radar stations on the coast, the RAF could coordinate its defence against this aerial threat, which proved to be of decisive importance in the Battle of Britain in 1940 (Terraine, 1988, 28–53).

Germany articulated the second vision. It was born to some extent out of desperation which stemmed from its poor geopolitical situation at the end of the First World War, where, as in the past, it was surrounded by hostile states. Drawing on the country's historical experience (in particular, Frederick the Great's campaigns during the Seven Years War and the tactical lessons it had learned from the First World War), the Germans sought to exploit their interior lines of operation by increasing the speed and mobility of their army. As in the past, the strategic imperative was to avoid engaging in a prolonged war. Instead, Germany sought to develop the means to fight and win quickly and decisively. It aimed to achieve this goal by employing a combination of the tank, airpower, mobile artillery and improved radio communications to ensure commanders exercised a high degree of flexibility over their soldiers. The fusion of these elements and their tactical potency was demonstrated to the world in the form of the panzer division. This combined arms formation operated in conjunction with other panzer divisions and close air support provided by the Luftwaffe to achieve surprise, shock and dislocation. Through such action, this combination of force was designed to overpower a defensive system by concentrating force typically at the weakest point of an enemy defence and rupturing the enemy line of defence by using firepower and the speed of its panzers to pour into a breach and then break out. Once through the enemy's defence, these forces could encircle and destroy the trapped enemy. The exploitation of radio communications was of fundamental importance in the realization of the German vision of future war in the interwar period. The widespread use of radio sets in tanks facilitated faster decision–making and greater flexibility regarding how tanks were employed in battle. This proved an essential advantage to the Germans in defeating the French and British, whose armies possessed more tanks but lacked effective command and control. It also provided a vital capability for facilitating air–land integration and the provision of close air support in battle, a capability also lacking in the French military in 1940 and which the Germans exploited ruthlessly to precipitate the moral collapse of the French army in the area of Sedan in 1940 (Guderian, 2001).

The third vision of future war that emerged during the interwar period was a synthesis of the two paradigms described above. This image was crafted by Soviet military theorists reflecting on the Russo-Japanese War, the First World War and the Russian Civil War. Like the French, its starting point was a deeply held belief that the next war would be total, requiring the mass mobilization of the state and society to wage it. It also recognized that technology reinforced the power of the defence and improved range, accuracy and rate of fire delivered from air and ground forces reinforced this tactical superiority (Triandafilov, 1994). However, like the Germans, the Soviets remained deeply committed to an offensive form of war. This reflected their view that the Soviet Union faced implacable enemies bent on its destruction. How then could they avoid repeating the experiences of the First World War? Like the Germans, the answer lay in the development of technological, tactical and organizational changes intended to overcome any defence's power. On paper, the Soviet answer looked very similar to blitzkrieg. However, the important difference was its connection to total rather than limited war, which implied that a quick and decisive victory was not possible within the broader setting of modern war (Vlakancic, 1992).

The Second World War relied heavily on the use of weapons mentioned above to wage modern war based on the principles of mass. However, the combination of mass armies with these weapons and capabilities did not repeat the large-scale stalemate and siege warfare that characterized and defined the First World War. Despite the belligerents' efforts to establish a superior defence, the enemy invariably crafted the means to ensure this remained a temporary phenomenon in campaigns. Within this context, command technology and with it the state's role in military, and scientific research, was firmly established as an integral part of war.

However, it is important to note that the technological experiences of the belligerents produced very different outcomes. In the case of Germany, the resulting power imbalance between the Axis powers and the Allies proved enormous, and technology provided one possible way to compensate for this differential in material power. The Allies had a pronounced advantage in terms of population: 360 million versus 195 million. The Allies also accounted for over 60 per cent of the world's industrial output compared to 17 per cent for the Axis powers; the entry of the US into the war was decisive as it accounted for 40 per cent of global arms production.

Germany never had enough tanks, aircraft and artillery to replace the losses it suffered during the latter stages of the war. It lacked the productive capacity to compete with the Allies, as demonstrated in Table 3.1. Recognition that it could not compete against its numerically superior adversaries caused Germany's political leadership to turn to technology to provide it with a silver bullet that would enable it to extract itself from defeat. As a result, Hitler sought to address the strategic and operational dilemma facing Germany by inventing new 'wonder weapons' that he hoped would allow it to deliver a knockout blow against its enemies (Kershaw, 2000, Chapter 13). In 1942 Hitler's cronies saw massive potential in developing the work of the scientific community's research on rockets. As a result, developing what became known as the V1 and the V2 programmes became a priority. Both were technically immature, but the state supported the first launch of the V1 which took place in December 1942. This and subsequent test flights highlighted persistent problems regarding the guidance system's stability and accuracy. It was not

Table 3.1: Comparison of armaments production between Allied and Axis countries during the Second World War

Weapons/Equipment	Allies	Axis
Tanks	227,000	52,000
Artillery	915,000	180,000
Mortars	658,000	73,000
Machine guns	4,744,000	674,000
Trucks	3,000,000	595,000
Aircraft	417,000	146,000
Transport aircraft	43,000	4,900

Source: Based on Overy (1995).

until May 1943 before the V1 was deemed to be technically reliable. Large-scale production began in January 1944, and the infrastructure to launch the V1 was built in the Pas De Calais in north-eastern France. This location ensured London was in the range of these rockets. It was hoped that 1,500 V1s would be launched against London over the first ten days, and it was anticipated that within three weeks, Britain would surrender. The success of the Allied landings on the Normandy beaches on 6 June 1944 caused Hitler to place even greater faith in the V1 as a war-winning weapon. In total, 8,167 V1s were launched against London over three months (13 June until 1 September 1944) killing 5,000 and injuring 16,000 civilians. However, this did not cause the moral collapse of the British as had been hoped. The initial panic caused by the V1 dissipated quickly as the Allies' actions reduced the number of missiles launched from France to a point where air defences could cope with the remains of the barrage, which meant fewer and fewer bombs got through. This onslaught against London finally ended with the ejection of the Germans from France, which meant they no longer had bases within range of Britain's capital. The V2 ballistic missile now came into its own and the first of these was launched against London on 8 September 1944. Some 2050 of these missiles were fired, and the last hit London on 27 March 1945. In total, this weapon killed 2,724 and injured another 6,500 people. According to the military historian, Max Hastings, 'Hitler made an important mistake by wasting massive resources on his secret weapons programme. The V1 and subsequent V2 rockets were marvels of technology by the day's standards, but their guidance was imprecise, and their warheads too small to alter the strategic outcomes' (2009, loc 8008). It is claimed that the combined cost of these programmes equated to the investment the Americans made in the Manhattan Project, which led to the creation of the atomic bomb.

Given the importance placed on weapons to solve Germany's strategic challenge of avoiding catastrophic defeat, why did the Nazis fail to develop their own atomic bomb? According to Weinberg, 1995, the failure to invest in this capability stemmed from a miscalculation about how much uranium 235, the best material to be used in the fission process, was required. This isotype is extremely rare, and very difficult to extract, and the German scientific community believed a large quantity was needed to create a sufficiently powerful bomb. This error led to the conclusion that developing such a weapon would take many years. Moreover, at the start of the war, such an investment was unjustified because the German state demonstrated an ability to wage short, decisive wars. By the time it became clear Germany was losing the war, it was too late to start the development of nuclear weapons because defeat was on the horizon.

It is important to note that it was not just the Nazis who saw technology as a potential war winner; so too did the Americans and the British. However, they adopted two research strategies that proved effective. The first focused on developing more efficient forms of manufacture, which meant controlling technological novelty in proposed weapons projects and concentrating on the best way to mass produce this capability (Hartcup, 1993, xxii). This imposed a critical constraint on the level of innovation incorporated into weaponry used at the forward edge of battle. Guy Hartcup noted in the desperate circumstances of the Second World War, the need to get weaponry into service as quickly as possible imposed a practical limit on the amount of research and development that could be conducted. Indeed, the primary concern of weapons development during the Second World War was the need to shorten the procurement cycle and get equipment into service as quickly as possible. In the case of the British, for example, they adopted the maxim 'second best tomorrow'. Such an approach entailed the partial completion of a weapon in terms of its development to a point where it was useable, moving to production and refining its capabilities, considering recent operational experience (Hartcup, 1993, 24). Today we call this process innovation, which has played a critical role in making new technologies available to the masses at an affordable cost.

Overy supports the idea that the Allies defeated the Axis powers partly because of their 'genius for mass production' (1995, 180). This, it is claimed, was a conscious decision made by Allied governments to overwhelm their enemies. In the case of the Soviet Union, the state played an instrumental role in ensuring industry continued to produce vast quantities of material even though it had lost so much of its natural resources. In the opening phase of the war, the Germans seized control of half the Soviet Union's grain supply, and meat production also fell by half. Iron, coal and steel production was cut by three-quarters, and the availability of vital minerals in arms production, including aluminium, manganese and copper, fell by two-thirds (Overy, 1995, 183). The success of the Soviet production system was demonstrated by how well it coped with the lack of resources available. As Overy demonstrates, in 1943, the Soviets used 8 million tons of steel and 90 million tons of coal to manufacture 48,000 heavy artillery pieces and 24,000 tanks.

In contrast, in the same year, Germany used over four times as much steel and over three times as much coal to produce 27,000 pieces of artillery and 17,000 tanks (Overy, 1995, 182). Much of this success was attributed to the three previous five-year plans beginning in 1928, which had been used to facilitate the rapid industrialization of the Soviet Union. This provided the foundation for its resilience and success in the Second World War.

The US was already the world's largest manufacturing economy by the start of the war. Its great success was converting a largely consumer-based economy into an enormous military–industrial enterprise. As Overy highlights 'where every other major state took four or five years to develop a sizeable military economy, it took America a year' (1995, 192). The key to success here was the partnership between the government and big business, which could deploy its technical and organizational skills to apply mass production techniques to a range of new products required by the military.

However, like the Germans, the British and the Americans also became more heavily immersed in science and technology during the war and increasingly looked to this domain to address some of the more challenging problems they confronted in destroying their enemy's military capability. The refinement and creation of new weapons technologies became the second principal strand of their research strategy. The rising importance of science in war can be seen in the dramatic increase in the number of scientists employed by the armed services. For example, during the war, the British Admiralty's scientific staff increased from 40 scientists and technicians to over 40,000. This vast increase in brainpower stemmed from the growing effectiveness of Germany's naval blockade of the UK. With the fall of Norway and Denmark in Spring 1940 and then France and the Low Countries in the summer of that year, the Germans were able to exploit their control of this expansive coastline and its ports from which to attack the UK's critical vulnerability – its reliance on trade to survive. The British had to import two-thirds of its food, 30 per cent of its iron ore, 80 per cent of its soft timber, 95 per cent of its petroleum products and 100 per cent of its rubber and chrome (Kershaw, 2000, 352–4).

Between 1939 and 1941, the British lost over 2,000 merchant ships to a combination of submarines, merchant raiders and aircraft. In 1938, the UK imported 68 million imperial tons of goods, but in 1941 this fell to 26 million tons. Most importantly, it lost more ships than it could replace and in 1940 alone lost 25 per cent of its merchant fleet. Science and technology played a vital role in shaping the outcome of this war, what became known as the battle of the Atlantic, an operation that also drew heavily on the RAF and USAAF to provide long-range patrol aircraft to operate over this vast expanse.

In the case of the war at sea, two vital enablers were intelligence that focused on deciphering enemy encoded messages to determine where German U boats were operating, and the development of better capabilities to wage this war. In the case of intelligence, the challenge facing UK code breakers was how to break codes fast enough to protect Allied shipping. Alan Turing's bombe, in crude terms, squeezed the equivalent of 12 Enigma machines with their coded wheels into one system that could then go through all the billions of possible options for each letter before deciphering the communication in time fast enough for it to be used as actionable intelligence. In terms of the war at sea, technological innovation included the relatively low-tech introduction of the Leigh light, a powerful searchlight that helped identify the location of German U boats moving on the surface at night when they recharged their batteries. More challenging was the development of a more powerful and sensitive radar that could detect a submerged submarine's periscope! In addition, new weapons and munitions were developed to address the specific vulnerabilities of German submarines. Innovation of this kind happened with improved tactics to counter the threat posed by U boats and better resourcing for air and sea operations in the Atlantic, including new bombers and aircraft carriers with aircraft capable of conducting anti-submarine warfare. As a result of these changes, only 57 Allied ships were sunk by German submarines in 1944, representing only 3 per cent of the losses suffered in 1942 (Overy, 1995, 177).

The development of new technologies also played a critical role in helping strategic bombing realize the hopes of those who predicted in the interwar period it could be a cheap and easy way to win a war. However, contrary to their expectations, bombing the enemy homeland did not result in the collapse of the nation's morale and the end the war in a matter of days. The British bombed the German homeland for the better part of five years and the results of its efforts are still contested to this day. Somewhere between 12 and 25 per cent of a bombing force was lost on each mission before effective countermeasures were devised. The efficiency of large-scale bombing was hampered by the effectiveness of Germany's air defence system, which was improved when an integrated air defence system was created over northwest Europe and Germany in 1941. This linked ground-based radar with anti-aircraft artillery (AAA) and fighters to engage enemy bombers. Over 55,000 AAA were deployed over the main route followed by British and American bombers and by 1944 over half the Luftwaffe was committed to defence of the homeland. But even before this integrated air defence system was operational, the British struggled. In 1940–41 they were unable to replace the number of bombers lost to Germany's air defence. It was also discovered that bombing was incredibly inaccurate, and most bombs failed to land within two miles of the target. In their efforts to improve the effectiveness of strategic bombing, the Allies devised new ways to jam or confuse the enemy's ground-based radars. In 1942, the British introduced a radar jammer called Mandrel, which emitted a solid signal to blind enemy radar. A technically more straightforward solution was the use of thin aluminium strips, which were dropped in large numbers to confuse the enemy radar making it impossible to distinguish the white noise of the window from the aircraft (Price, 1967, 28-53).

The stakes in the war in the electronic spectrum increased further with the development of radio beams used to navigate aircraft. This capability was initially developed to help civilian aircraft navigate to airfields during poor weather. It was applied to strategic bombing during the Second World War and allowed bombers to navigate toward their targets at night. In Operation Totalize, a similar navigation method was also used to help British ground forces launch a large-scale night offensive against the Germans. The reaction to these technological innovations was a war within the electromagnetic spectrum.

Most significant in the war–state relationship was how this partnership pushed the boundaries of existing knowledge. As Zuckerman explained:

From the time of the industrial revolution until the Second World War the technological needs of the armed forces were generally met out of the same scientific and technical knowledge with manufacturing industry put to use in satisfying commercial demands. The picture became transformed when … scientists instead of standing back and allowing the military to get on with their own affairs, stepped into the arena with proposals such as those which led to the development of radar and the nuclear bomb. (Zuckerman, 1966, 28–9)

The idea of the Second World War being a watershed in terms of military research was shared by Pearton:

The destruction of Hiroshima and Nagasaki marked a turning point in the preparations of war. Nuclear physics and the associated engineering were not the only spheres of the unknown which had to be explored. The use of the jet engine for aviation and experiments with rocketry took warfare into new environments, the nature of which was largely unknown. … In consequence, basic research entered into the production of even conventional weapons to a degree which had not been previously experienced. (Pearton, 1982, 246)

The dropping of the atomic bomb represented the apotheosis of what command technology could do if the state and the private sector worked in partnership on a project that was not limited by resources. In this case the development of the weapon cost over US$2 billion and required a workforce of 120,000 scientists and engineers. Its success and the fruits of the revolution in rockets and guidance technologies produced in Germany and computers developed to break German encryption combined to produce a suite of weapons that signalled profound change in the character of future war but which helped entrench the war–state relationship in the post-war world. However, before exploring the new scientific and geopolitical context created by Allied victory in the Second World War, it is important to reflect on how technological change shaped industrialized warfare.

Less spectacular but more critical in terms of the day-to-day lives of people in the future was the development of the computer Colossus by scientists and engineers at Bletchley Park and the research division of the General Post Office. History tells us that ENIAC was the first programmable electronic computer, but Colossus, also a programmable computer, carried out its first operation on 2 February 1944, many months before ENIAC became operational. Its principal function was to speed up the deciphering process for German communications encoded on their Enigma machines and was created in the hope it would provide the Allies with a significant intelligence advantage before they committed their forces to the invasion of Europe in June 1944. The principal advantage of this machine was its speed. The Mark 2 version was capable of processing 25,000 characters per second and exploring the billions of different settings on Enigma before breaking the code and providing the message contained within. Colossus deciphered over 63 million text characters in the last year of the war. This vital intelligence showed the deployment of German forces and their intent and helped Allied planning. Such was the importance of this information that it was claimed that its use shortened the war by two years. After the war, it is believed that Alan Turing and Max Newman, two of the key figures who helped break the Enigma codes, took parts of Colossus back to their respective universities and the work they did at Bletchley fed directly into the UK's research on computing in the 1950s and 1960s.

Conclusion: The concept of modern war

The concept of modern war is defined by mass, and this can be linked directly to the industrial revolution, which allowed governments to capitalize on the exploitation of nationalism and democracy to provide the means to wage modern war. How then did these elements change the character of war? Strategically, it meant an unprecedented scale of resources was available to the state to bring about the rapid defeat of an enemy either in a limited or total war; in essence, war was no longer constrained as it had been. In narrower military terms, this typically meant generating the largest possible force in the shortest time. These vast armies, which by the time of the Franco-Prussian War (1870–71) consisted of a million troops, could be sustained, at least temporarily, by the wealth generated by industrialization. It is interesting to note that global economic growth had been broadly flat over the preceding 2,000 years – the global economy grew on average by approximately 6 per cent per century – but it increased dramatically during the industrial revolution. The revolution also meant that states possessed the capacity to provide the supplies needed to sustain these vast armies.

Most importantly, states could field forces for extended periods and were no longer fixed by the seasons as they had been in the past. The biggest fear was not the breakdown of supply and the disintegration of the army, which had been the principal concern for commanders in the past, but that the financial resilience of the state could only tolerate a supreme effort over a short period before it became exhausted and collapsed. This was one reason military planning in the lead-up to the First World War focused on winning in the first few months, and why their plans were plagued by over-optimism. Pre-1914, the idea that a war of this scale could drag on was simply inconceivable to the generals responsible for winning the next war. However, they underestimated the capacity of the state to endure and conscript the resources needed to fight over an extended period. As a result, military campaigns were waged all year, even amid the worst of winters, and, as the two world wars demonstrated, war could be waged at a scale and intensity that ran for many years. To dig into the human resource of war in this way also required modern war to tap into the political revolution associated with the Napoleonic Wars. The legacy here was not the democratic revolution but rather the legacy of nationalism which was evoked by governments to legitimize universal military service and facilitated the supply of human resources required to wage the total wars of the 20th century. It is asserted that this political mass mobilization was mainly possible because of new communications technologies in the form of mass media, principally the press and eventually radio. However, the role played by political factors that facilitated the exploitation of technology is more important.

Mass industrialized warfare and the challenges of defeating the enemy's vast forces resulted in a new level of war. In the past, the conduct of war was divided into two levels: the conduct of battle, which was defined as the tactical level of war; and strategy, which was the exploitation and battles to achieve the political aim of the war. The emergence of an operational level has been described as the creation of a gearing mechanism that coordinated the movement of mass formations in a campaign theatre so that battles worked towards achieving the strategic goal. As such, it recognized that, in modern war, victory and defeat were not likely to be the consequence of a single battle but many battles happening simultaneously in time and space. Inevitably, the industrial revolution significantly impacted how military campaigns were conducted. The railways provided commanders with a fast and efficient means to transport armies over long distances, which allowed for the rapid concentration of force within a theatre of operations. The railways also acted as vital arteries which provided the supplies required to keep armies fed and armed as the campaign progressed. The principal outcome of this new form of transport was to dramatically increase the speed and distance over which armies could operate and removed an important constraint because logistics or the provision of supplies could be hauled by train rather than vast numbers of wagons pulled by animals, which meant space also had important implications for supply as sufficient fodder also had to be carried to keep pack horses alive, which reduced available supply for the army. It also limited the range of an army, because at a certain distance the animals would have eaten all the supplies they carried. This, then, made it necessary for armies to stop and resupply at frequent intervals, which slowed an army's progress and the distance it could cover during a campaign season.

The mobility of armies was enhanced further due to the invention of the automobile. This facilitated the movement of armies between the railhead and the front line. The motorization of logistics and the development of specialized armoured forces combined allow rapid battlefield movement and create opportunities to manoeuvre into a position of advantage. Usually, this allowed a force to seize key terrain or attack the enemy's rear area and supply routes and headquarters. In the past, winning relied mainly on what happened on the actual battlefield, but the increased mobility of armies now allowed new ways to strike at the enemy and bring about their defeat.

Good command and control were vital if the movement of armies within a theatre was to be orchestrated to defeat the enemy. Technology also played a vital role in ensuring the commander was able to impose their vision of how the plan should unfold. Initially, this was achieved via the telegraph. This was used significantly to organize the rapid mobilization of armies in the lead-up to war and the deployment and line of march taken by higher level formations once moving in the theatre. Introducing a wireless radio in the Second World War enhanced command and control and permitted the dispersal of forces across a geopolitical space. Given the scale and complexity of moving mass armies towards their objective, which was to find and give battle, improvements in command and control were vital if coordination was to be achieved before the battle happened.

Again, however, the effectiveness of technological change was amplified by other factors. In this case, it was organizational and structural reforms that were not directly attributable to the industrial revolution or the changes it had wrought in the conduct of war. In the aftermath of the Seven Years War (1756–63), the French army, which the armies of Frederick the Great had humbled, introduced a series of organizational changes designed to improve its tactical flexibility. This entailed creating a new formation called the division, which consisted of approximately 10,000 men organized in a group containing artillery, infantry and cavalry. Three divisions were then organized into corps (30,000 troops) and three corps made an army (100,000 troops). Twentieth-century armies became so big that it was possible to have many armies operating in a theatre. When this happened the need to simplify command resulted in two or more armies being combined to make an army group. The expansion of the commander's control coincided with the increase in the size of armies. By the time of the Austro-Prussian War (1866), it became impossible to concentrate a nation's armies in a single space. Consequently, the organizational structure which had its start in the 18th century and was firmly established during the Napoleonic Wars provided the means to allow forces to march dispersed across a range of routes but to concentrate when necessary.

Technology also revolutionized the face of battle. The principal cause of this revolution was the dramatic increase in firepower measured in terms of the increased rate of fire, accuracy and extended range of weapons. In the case of artillery, this resulted in guns firing shells over a mile or more, depending on the calibre of the gun, and the development of new forms of targeting as the enemy could not be seen behind the frontline of battle. Of importance here was the role played by airpower in providing the means to see behind the enemy's front line. Technical improvements in airpower, rifles and artillery, and the machine gun increased the depth of battle space determined by the range of weapons. This complemented the increase in battle frontage as armies grew. In addition, to reduce the casualties inflicted by the storm of metal created by this firepower revolution, forces became more dispersed across the front line and increased the battle space's size. This was then compounded by the use of outflanking to avoid direct assault, which extended the frontline further.

In contrast to the Napoleonic Wars (1793–1815) or the American Civil War (1861–65), the Franco-Prussian War revealed the potential nightmare that was to unfold on the battlefields of the First World War. By now, armies were so large that multiple rather than single battles were happening simultaneously. As each commander sought to outflank their opponent rather than charge their frontline directly, these engagements began to link. This trend happened again during the First World War, resulting in separate engagements becoming connected to create one massive battle across the entire front. Most importantly, with no exposed flanks, armies were now forced to conduct frontal assaults. To avoid the effects of persistent enemy fire, armies dug in and entrenched. The result was the emergence of trench deadlock. This pattern was demonstrated in the Russo-Japanese War (1905) and was developed to its highest form during the First World War. One of the salient aspects of modern war was the realization that the new technological and material conditions made the pursuit of decisive battle all but impossible, but this continued to be the main goal that shaped the conduct of military strategy and operations. It is claimed that the resulting dissonance between strategy and tactics was not resolved until after the First World War. Based on the reflections of largely Soviet military theorists in the interwar period, it became clear that modern war was characterized and shaped by the power of the defence. The challenge was to devise a means to overcome this tactical matrix. Again, technology is perceived to have played a critical role in achieving this goal in the form of the tank and its combination with artillery, infantry and airpower, which came to be known as blitzkrieg warfare. The memory of war always allocates the central role to the ending of trench deadlock in the First World War to the tank, but this is misleading. The German Michael offensive in 1918, which destroyed the British 5th army and broke through Allied lines, was largely caused by a tactical reorganization of their infantry using infiltration tactics combined with hurricane bombardments and airpower. In essence, the exploitation of technology in war relied on an adaptation of forces to operate in this new setting captured in tactical military doctrine.

The Second World War fulfilled the prophecies made in the interwar period. It was essentially a repeat of the first regarding the material conditions that shaped operations and battle. The critical difference was the expansion of the battlefield caused principally by the role of airpower which was critical to the success of ground operations, and the use of armoured forces by all sides to break down the enemy's defence and then exploit deep into the enemy's rear area. Perhaps the apotheosis of modern military operations on land was Operation Bagration. Launched on 22 June 1944, it entailed an offensive of 2.2 million men attacking on a front of several hundred kilometres and resulted in the destruction of a German army group consisting of nearly a million soldiers and forced them to retreat 750 km with their new frontline based on Warsaw and the Vistula. Although seen as the worst defeat inflicted on the German army, Hitler retained sufficient force to continue fighting for another nine months before the Soviets finally occupied Berlin in April–May 1945.

The two world wars of the twentieth century illustrate how the industrial revolution impacted on the conduct of war. Most important was the rising importance of science and technology in war and the role of the state as generator of this capability. This marks the high point of the war state relationship. The next chapter explores how the relationship between war, technology and the state evolved as governments grappled with the challenge posed by the advent of the nuclear age.

4

The Nuclear Revolution and
the Rise of Postmodern War

This chapter explores how the nuclear revolution in war changed the character of conflict and challenged our traditional conception of what constituted an act of war. The related question of how this change impacted the war–state relationship is also explored. I explained in the previous chapter how the rise of modern war is intimately connected with the industrial revolution. Innovations in this period transformed warfare and contained several military revolutions on land and the sea. One of the most pronounced of these 'revolutions' was the emergence of airpower as a distinct but separate environment of modern war. The rise of electronic warfare was less well known but as important. However, in terms of impact, dropping two atomic bombs on Japan in August 1945 represented a true paradigm shift in the conduct of war. It is often pointed out that the destruction of Hiroshima and Nagasaki was genuinely awful if measured in terms of deaths, but the USAAF firebombing of Tokyo in the preceding March was far worse; between 80,000–100,000 killed versus 66,000 at Hiroshima and 39,000 at Nagasaki. Viewed in this way, the explosive potential of the atomic bomb represented a critical enhancement in the firepower available to the US but, at the time, it was not necessarily seen as a game changer in the conduct of war. It is important to note that the significance of the atomic bomb in terms of its conduct on war is contested. In the view of Price and Tannenworld (1996), this weapon was viewed as just another weapon, albeit one that packed a lot more punch, to be employed on the battlefield. As such, its claimed impact has been much exaggerated. In their view, the most crucial change was the development of thermonuclear weapons. The detonation of the first hydrogen bomb in 1952 produced a ten-megaton explosion approximately 500 times more powerful than Hiroshima. What possible use could such a weapon have but to prevent war?

The subsequent introduction of intercontinental ballistic missiles, heralded by the Soviet Union's launch of Sputnik, gave this weapon an almost invincible aura.

However, as Freedman explains, the significance of this technology lay in the fact that a single bomber armed with a single bomb was able to destroy a city (Freedman, 2003, 17). This marked a dramatic increase in the efficiency of industrialized slaughter and compared favourably to the huge investment of nearly 300 B 29 Super fortresses required for the firebombing of Tokyo. The implications of this dramatic increase in both the speed and scale of destruction represented a profound challenge to the conduct of mass industrialized war. This was particularly true once the Soviet Union acquired the technology to build its atomic arsenal in 1949. In this new geopolitical setting, deploying mass armies, naval armadas or aircraft fleets in a war of mass attrition no longer made military sense. The destructive power of a few atomic weapons would annihilate any concentration of military force. As significant, the speed and scale of destruction made the mobilization of industry and the population pointless because it could not instantly replace such losses on the battlefield. As a result, ironically, a weapon that came to represent the perfected logic of modern war became the principal source of its demise in the Western world.

The existence of nuclear weapons constrained the traditional Western way of war, especially its supposed passion for battle, which was viewed with increasing horror by the likes of Keegan, who concluded in his book *The Face of Battle*, that the modern battlefield had evolved into something humanity could no longer endure or cope with. It also changed the state's relationship with society as the traditional model of total war became problematic within the context of a nuclear conflict which was likely to result in rapid and widespread destruction. War in this new context was described as a 'come as you are party', in effect you fought with what you had and little thought was given to subsequent mobilization. So the role of the citizen soldier, an essential aspect of the Western way of war school, also declined in importance, causing the military to become increasingly removed from the day-to-day experience of society. In contrast to the period leading up to the First World War, people during the Cold War genuinely feared the outbreak of a general war simply because its potential consequences were apocalyptic.

Quite simply, war was no longer seen as a rational instrument of policy. In his book *Retreat from Doomsday*, John Mueller went so far as to argue that this cultural shift and not the presence of nuclear weapons explained the absence of a war between the great powers during the Cold War (1989, 42). In this case the horrors of the two previous world wars and the destruction it inflicted on the principal belligerents was sufficient to destroy any appetite for war. These two positions were not mutually exclusive and reinforced each other in that, having experienced the horror of war and seeing the vast destructive capacity of nuclear weapons, most ordinary people and their governments shared a genuine anxiety about the outbreak of a Third World War. The industrial revolution also changed the balance sheet of profit and loss. Total war absorbed an unprecedented level of investment in constructing a viable war machine capable of defeating other major industrialized states. The wanton destruction of physical capital was intended to break the opponent's ability to wage modern war and, finally, the wealth lost in terms of the lost economic growth which resulted from fighting ensured that war made little sense in economic terms.

It is essential to recognize that the nuclear age changed many aspects of the connection between war society and the government, including its relationship with the state. However, this did not necessarily imply the connection between them was significantly weakened, as has been assumed (van Creveld, 1999; Bobbitt, 2002; Mann, 2013). I contend that a more nuanced connection between war and the state emerged in the nuclear age, one that reflected the needs of the state during this period, and which ensured it remained a vital pillar in the creation and employment of military power. Paradoxically, the advent of the nuclear age created a compelling logic which meant that even though war was too terrible to contemplate, to ensure this did not happen required that governments prepare for the worst and demonstrate they had both the capacity and will to fight a nuclear war – 'if you want peace, prepare for war' (Vegetius). This requirement ensured the state continued to play a central role in providing the means required for the military to deter a general war during the Cold War. Of importance here is not whether deterrence worked or not, but instead the actions carried out by governments and state bureaucracies which focused on the most elaborate thinking and preparation for a future war. It is also important to acknowledge that enforced peace in Europe did not extend to other regions of the world; the logic of the Cold War inflamed instability and violence across the globe and Western military power often played a role in these wars. Described as 'limited wars', brushfire wars' or wars of 'savage peace', these conflicts took place largely in the developing world and while facilitating dramatic political change they entailed a huge human and economic cost. This, too, needs to be taken into account when thinking about the war–state relationship. My central argument is that the West's response to both the challenges posed by the Cold War and the many smaller hot wars that erupted over this period caused Western governments to rely on technology to provide security and stability, and the state played a vital role in facilitating innovation which implied a significant role in accelerating the third and indirectly the fourth industrial revolutions. Two crucial questions emerge from this overview. First, we need to explain why technology moved to the centre of defence planning. Second, we need to understand the implications of this drive for innovation in terms of the war–state relationship.

The impact of the nuclear revolution on war

The nuclear age precipitated a profound change in the organization and conduct of war. Hables Gray (2013, 22) asserts that post-1945 marks the dividing line between modern war and the birth of what he terms postmodern war. This philosophical construct is used as intended, not as a label, but as a way of demonstrating how war, like many forms of human activity, is a discourse. That discourse changed profoundly because, by 1945, scientific advance, in the form of nuclear weapons, made modern war between nuclear-armed powers all but impossible so long as they possessed the means to respond to an initial nuclear strike. As Brodie (2014, 205) observed in this new setting: 'Thus far the chief purpose of our military establishment has been to win wars. From now on its chief purpose must be to avert them. It can have almost no other purpose.'

This new strategic setting precipitated what Holsti (1991) described as the diversification of warfare. This resulted in a blurring between peace and war as governments employed various means to achieve their policy goals below the threshold of general war. Most importantly, the forms of war proliferated as new ways were devised to employ war as a political tool in a nuclear world (1991, 270–71). This change did not make Clausewitz's concept of war obsolete but required it to be adapted (Cimbala, 2012).

Clausewitz explained, 'war is an act of violence to compel our opponent to fulfil our will' (1976, 77). War is also the continuation of policy by other means (1976, 82). War then is defined as a discourse of physical violence to achieve a political goal. However, in examining the war–state relationship in the West post-1945, it is essential to revise our understanding of war and recognize that it became necessary to wage war without physically destroying the enemy in the nuclear age. Russian military reflections on the Cold War reveal an interesting narrative that reinforces this view. According to this analysis, the Soviet Union lost the Cold War because it was defeated by non-military means employed by its enemy that focused on psychological, political, information, social and economic attacks against the Soviet State (Fridman, 2018, 49). Technology played a vital role in facilitating this defeat via the communications revolution. However, the most salient aspect of the Cold War was the discourse of deterrence. Within this context, the rituals of war in terms of organizing, preparing and demonstrating an ability to fight a nuclear war in the hope of deterring and preventing the possibility of war became substitutes for organized violence. Small wars happened on the periphery of the United States and Soviet geopolitical space, but in the core region, a different kind of cognitive and cultural violence emerged, which can be seen as a form of war, (Cramer, 2006, 1–20).

How then did technology fit into this new discourse of war? According to Buzan (1987, 216), because nuclear deterrence relied on anticipated weapons performance, it became sensitive to technical innovation, which meant the state had to respond to technological change by investing in defence research to maintain the credibility of its deterrent. The priority was to ensure the state possessed the capability to withstand a surprise nuclear attack and have enough nuclear weapons of its own to inflict unacceptable damage on the aggressor state's homeland. This partly explains why both superpowers acquired thousands of nuclear weapons and why they quickly escalated from atomic bombs in the kiloton range to hydrogen bombs, which produced an explosive capacity that equated to millions of tons of TNT. It also explains why they invested in multiple delivery systems. In the case of the US, it maintained a nuclear triad based on strategic bombers, intercontinental ballistic missiles and eventually submarine-launched ballistic missiles. Possessing multiple means to counter a first strike induced certain strategic stability as it reduced the risk of a surprise attack against the enemy's nuclear arsenal. The advent of anti-ballistic missile defence (ABM) systems challenged this delicate balance but was then countered by the technological innovation of placing multiple warheads on each missile, which it was believed would overwhelm an ABM defence – only one warhead was needed to destroy the target and hence the credibility of the deterrent was maintained. Why then, having reached this moment of stalemate captured in the idea of mutually assured destruction, did the West, in particular, continue to seek to push against the existing frontiers of military technology? This is particularly important in view of the costs incurred by both the US and the UK. The diversion of scarce resources into defence research was perceived to play a role in the declining economic fortunes of both countries, especially when compared to the economic miracles of Japan and West Germany, both of which preferred to invest public money in the development of civil technologies which provided a better economic return. This problem was particularly acute in the case of the UK, which seemed to be trapped in the economic doldrums in the 1970s (Chalmers, 1980).

Friedberg notes that throughout the Cold War, the US devoted a huge share of its wealth to achieving and maintaining a pronounced advantage over the Soviet Union in most areas of military technology. He believes this pursuit of military technological excellence was prompted, in part, by a belief that the free world was at a disadvantage in terms of its defence because, in contrast to free market economies, command economic systems were not constrained in the extraction of resources which were then used to support defence, especially in peacetime. It was also believed that Communist societies were less sensitive to the issue of casualties. This disparity was evident in the proportion of wealth spent on defence by the Soviet Union, which was over 15 per cent of GNP, compared to 8 per cent in the United States. The Soviets also maintained the largest army in the world (4–5 million men) during this period. The belief that the 'superior extractive capacities of Communist states gave them certain inherent advantages in mobilizing military power led the West to rely increasingly on technology to act as a substitute for mass' (Friedberg, 2000, 208).

By the late 1960s, the Soviets achieved sufficient parity with the US regarding the nuclear balance, which produced a strategic dilemma for the Americans. In the event of a war in Europe, there was genuine concern that NATO's small conventional force could not withstand an onslaught from Warsaw Pact forces which outnumbered NATO in manpower and every category of weaponry. In the event of a war, it was believed NATO commanders would call for the use of tactical and theatre nuclear weapons within Europe in the first few days of the conflict. The United States' big fear was what to do if the Soviets failed to recognize the importance of these thresholds and continued to advance across West Germany. Faced with this situation, would the US government sanction attacks against the Soviet homeland if it knew they would reciprocate in kind? Was a president willing to sacrifice US cities to save their European allies?

This strategic dilemma caused policy makers within the US to devise a way of creating a conventional capability across NATO, making it possible to fight and contain military aggression at the conventional level. As a result, from the late 1960s onwards, both the United States and NATO increasingly relied on conventional forces to deter Soviet aggression (Freedman, 2003). However, during the late 1970s, the Soviets began reinforcing their numerical superiority by improving the quality of their forces as well. This presented a significant challenge to the West, which responded in kind, and NATO defence spending increased by 3 per cent per annum from 1977 onwards. This was intended to pay for the modernization of NATO forces and the incorporation of a new generation of smart conventional weaponry supported by a more elaborate digitized communications infrastructure. Fortunately, these weapons were never fired in anger during the Cold War but, by the 1980s, the acquisition of such complex weapons proved to be financially ruinous, with acquisition costs doubling every seven years and accounting for nearly 40 per cent of defence budgets (Kirkpatrick and Pugh, 1985, 59–80). As a result, we saw two problems emerge within defence. The first focused on the question of affordability. The second, caused by the first, was the response of governments when confronted by the challenge of rising costs. Within this setting, the answer was to buy less and compensate for the reduction in the size of the armed forces through qualitative improvements in NATO's conventional and nuclear arsenal. This meant focusing even more on technological performance to compensate for reduced numbers of aircraft, tanks and ships, but that then caused costs to increase, which resulted in a further reduction in the numbers of systems that could be deployed if war broke out. This phenomenon became known as structural disarmament (Smith, 1980) and confronted defence planners with a real dilemma in terms of achieving a balance between quality and quantity. The insanity of this behaviour was captured by the defence industrialist Norman Augustine (1997, 11), who predicted in the 1970s that by 2054 the entire US defence budget would be able to afford a single jet fighter.

War and technology beyond the Western world

In contrast to the decline in the incidence of war among Western powers was the increase in the frequency and lethality of war in the rest of the world after 1945. Of particular importance here was the dramatic rise in the number of civil wars. Based on an analysis of wars categorized as incurring at least a thousand battle deaths, Tilly notes that the number of new civil wars rose from about 10,000 battle deaths per year at the start of the 20th century to 100,000 per year from 1937 until the mid-1970s (Tilly, 1992, 201). Of importance here is the role Western powers played in these conflicts and the impact these wars had on the war–state relationship in the West.

Given the prominent role assigned to technology and war and the impact this had on war–state relationship in the Western world, it is perhaps not surprising that a major assertion made in this study is that the technological imperative also became a critical component in allowing Western forces to engage in wars with non-Western states. These wars took two forms. The first can be categorized as wars of decolonization caused principally by the rise of local nationalism. The outbreak of the Second World War and the defeat of colonial powers such as France, the Netherlands and British imperial interests in the Far East by the Japanese demonstrated the fragility of Europe's empires and, most importantly, the belief that colonialism was the inevitable result of the moral and material superiority of the West. In essence, their defeat by the Japanese destroyed the myth of their supremacy. Equally important was the role played by local insurgent groups in Japanese-occupied territories in areas like Malaysia and Vietnam during the Second World War. These groups were fighting for independence and not the return of European colonial powers to their countries. In addition, the declared ideals of the Allies as captured in the Atlantic Charter, which emerged in direct response to the horror of Nazism, recognized the inalienable rights of humanity, which was hard to reconcile with the legacy of empire. Having said that, at the end of the Second World War, European governments sought to restore their empires. However, even if the political legitimacy of empire was now contested, European governments remained determined to reclaim their former colonial territories, even if this was done in the face of increasingly hostile local resistance, which manifested itself as a series of wars across Asia and Africa.

Also of importance here were the changing attitudes of European citizens towards the legitimacy of preserving their colonies. The United States's formal position on the rejection of colonialism influenced the position taken by nations like the UK. In the case of other European governments, there is no doubt that the experience of Nazi occupation of France, Belgium and the Netherlands raised questions within their electorates about the morality of preserving their empires after the Second World War. This did not result in an immediate and dramatic shift in policy, and there were no protests when these states sought to re-establish control over their former colonial territories. However, once opposition emerged to the return of former colonial masters one suspects there was some sympathy for the cause of those seeking independence and most importantly a reluctance for their children to fight in these wars. The increased speed of communication and television and radio also opened a window on these wars that allowed European citizens to see in near real-time the brutality of the wars being waged on their behalf. This is cited as a defining moment in the changing balance of power between Europeans and local insurgents as increasingly the latter sought to exploit the media to bring the justness of their cause and their plight to the wider world. Although contested now it was believed the media played an important role in shaping the outcome of the US intervention in the Vietnam War (Summers, 2007).

The process of decolonization was complicated by two hugely important events which exacerbated instability and the incidence of war in the developing world. The first was the redrawing of the geopolitical map in the Middle East and the creation of the state of Israel, which was carved out of the British Palestine Mandate. This decision was imposed on the largely Arab population by the newly created United Nations and was perceived to be unjust by the Arab governments who sympathized with the plight of the Jewish population but did not understand why they should pay the price for Hitler's genocide. As a result, they challenged this political settlement and this resulted in a succession of large scale conventional wars, which did not end with peace but merely rumbled on into a series of unconventional conflicts. The second was the withdrawal of the British from south Asia and the partition of the British Raj into India and Pakistan. This was largely a consequence of inter communal strife between Hindus and Muslims but the hostility between these two states persisted after the creation of Pakistan because both states claimed the province of Kashmir, which was largely populated by Muslims and Islamabad believed it should be part of the Pakistani state. The Indian government rejected this claim and again the result was a series of conventional and unconventional wars.

The second strand of Western war with the non-Western world was a direct consequence of the Cold War. The nuclear stalemate imposed in Europe resulted in both the Soviet Union and the US seeking to project their struggle to other strategically important areas of the globe. This ideological skirmish between Communism and Capitalism was also connected to wars of decolonization in that local resistance often drew heavily on Communist ideology to legitimize their struggle to gain the support provided by other Communist states. As a result, the global conflict against communism often fused with colonial conflicts and resulted in these wars of national liberation being seen through the lens of the Cold War. This was why the US, which strongly opposed colonialism, supported the UK and France in their efforts to fight Communist insurgents in British-controlled Malaya and French Indo-China. The presence of nuclear weapons was largely irrelevant in this setting as the US discovered when the Chinese Communist Party defeated the Kuomintang in 1949, leading to the establishment of the People's Republic of China. The limitations of nuclear deterrence beyond the European geopolitical space were also revealed in 1950 when Communist North Korea invaded South Korea with the support of the Soviet Union. The presence of nuclear weapons also failed to deter Chinese intervention in this war when in October 1950, the US-led UN force pushed into North Korea, thus provoking a direct military response from China. As a result, the use of the nuclear arsenal had no role to play in addressing the challenge posed by what might be loosely described as Maoist revolutionary war. As Osgood (1957) explains, in an attempt to avoid watching successive countries succumb to the strategy of revolutionary war, the United States was forced to develop a limited war strategy that would allow it to challenge Communist insurgency via political, economic and military means, but in a manner that ensured a brush fire war in a remote part of the world did not escalate into a direct superpower conflict. This then provided the broad parameters that shaped the West's military interaction with the non-Western world. In sum, the character of war in the Cold War was marked by a pronounced shift in that we did not see a great power conflict between the superpowers, but rather proxy wars in which the United States and the Soviet Union exploited local conflicts to increase their influence and control in the developing world.

Why did the proliferation of small wars in this geopolitical space perpetuate the drive for technological innovation among Western militaries? This is a perplexing question because there is a general acknowledgement that best practice in terms of military strategy emphasizes the importance of labour-intensive rather than capital-intensive forces to defeat revolutionary warfare, which is characterized as being more about political action than finding and killing enemy forces, which is the principal function of military technology. This phenomenon has been largely explained by internal changes within Western societies after the Second World War. In the specific case of Western Europe, it has been asserted that a particular kind of war weariness infused national politics. In this instance, it was not fear of nuclear war that shaped people's attitudes to these small wars, but rather the realization that it was possible to achieve prosperity and enjoy a better life without occupying and colonizing the lands of other nations. This was clearly demonstrated by the economic success of West Germany and indeed even former colonial powers such as France and Italy enjoyed unprecedented prosperity and rising living standards during the Cold War.

As has been said, science and technology harnessed to power the industrial revolution was a hugely more efficient driver of wealth creation than colonialism. Post-war economic success in the West was paralleled by increasing concern over the wellbeing and opportunity of all citizens, and this resulted in the expansion of health and education and the drive to create an equal society. This, in turn, led to the emergence of an increasingly critical attitude to defence spending and even more the waging of war, which was reflected in the policy priorities of national governments. Within this new socioeconomic context the connection between the rights and entitlements of the citizen and the burden of military service diminished in importance (Tilly, 1992, 193–215). According to Yagil Levy (2012), this connection was partly broken during the Cold War because, in the nuclear age, the mass mobilization of millions of conscript soldiers made little sense. A different agenda now drove the advance of political freedoms, and the importance of war declined in terms of the electorate's wants and needs. Not only did war no longer pay, but also as important was the availability of many other ways in which the people could advance economically and politically. The existence of alternative sources of wealth creation made people increasingly sceptical that war served any meaningful purpose and people could envisage a prosperous life free of war. By contrast, serving in the military entailed a potentially high opportunity cost due to the risk of permanent injury or death (Levy, 2012, 534).

This asymmetry of interest between the Western military power and its local foe in a faraway conflict has been cited as a principal reason why, as Andrew Mack put it in 1975, 'big states lose small wars'. For those fighting for their independence, such conflicts represented an existential struggle. However, for the intervening power, whose homeland was largely safe from direct attack, there was often little at stake beyond prestige and honouring alliance commitments. Mack contends that this asymmetry in interest rather than poor generalship allowed the weaker power to prevail by exploiting the absence of strong support for war within the Western state. As he explains:

> Where the war is perceived as limited – because the opponent is weak and can pose no direct threat – the prosecution of the war does not take automatic primacy over other goals pursued by factions within the government, or bureaucracies or other groups pursuing interests which compete for state resources. (Mack, 1975, 183)

As a result, during the Cold War, conflicts in the developing world involving Western military intervention operated within a structural dynamic in which there was often no clearly defined payoff to justify the expenditure of national treasure and bloodshed for the Western force. This inevitably led to internal political division between those who supported the war for ideological reasons and those who perceived it to be futile and politically naive. This thesis has been contested by those who argue that Western military failure in non-Western wars since 1945 stemmed not from the asymmetry of interest between the belligerents, but mainly from the poor strategic choices exercised by commanders (Arreguin-Toft, 2001). Others have chosen to focus on the intrinsic weaknesses of democratic governments when embroiled in small wars of this kind (Merom, 2003). Finally, the third school of thought believes these wars were lost because of the heavy reliance placed on capital-intensive methods of war employed by the intervening power (Gibson, 1986; van Creveld, 1991; Lyall and Wilson, 2008). It is claimed that reliance on technology resulted in too great a fixation on killing insurgents, which caused innocent loss of life, and this defeated the goal of the war, which was to win the support of the local population within the conflict.

Typically, we attribute many of these problems to a fixation on the culture of the military and, in particular, its preference for large-scale conventional war, even in circumstances where it is not appropriate (Nagl, 2005). There are two problems with this argument. First, it fails to consider the variation between Western militaries in terms of their culture. For example, the UK has long prided itself on its proficiency in fighting small wars compared to its US counterparts. Second, Caverley (2009, 120) asks why if it is the military that decide to impose a mode of warfare that is inappropriate, their political masters do not overrule them and insist on a different strategy. He believes the problem is not so much the military but rather the political choices made by a democratic government in which capital-intensive techniques are consciously employed by national governments even though they recognize the risks involved. However, it is deemed the least bad option because the costs incurred are low; the alternative of focusing on a labour-intensive strategy represents a significantly more expensive option and exposes its citizens to greater risk. This implies that military organizations are constrained by the political choices made by their civilian masters and that even though they are aware of what is required to win or avoid defeat, their freedom of action is limited by the preferences of the electorate and the government. This at least provides a partial explanation for the reliance on technologically based solutions by Western states fighting limited wars.

Consequently, even though the nuclear age made modern war largely obsolete as articulated in the First and Second World Wars, the imperatives of nuclear deterrence and the proliferation of small wars that confronted the West during the Cold War ensured a sustained connection between war and the state. Most importantly, this relationship was shaped increasingly by the nature of technological innovation driven by the factors outlined above. The Cold War and the rise of postmodern war required a different kind of political economy, one which placed less importance on mass production and the generation of huge quantities of material to sustain large-scale war, to one which measured success in terms of ambitious technological leaps as Western states sought to preserve their technological edge against both state and non-state actors. As has been said in Chapter 3, the idea of qualitative arms racing – a military competition based on the performance of weapons rather than the size of a state's forces – first appeared on a significant scale during the second industrial revolution (van Creveld, 1989, 224). The emphasis on performance steadily increased over time, facilitating an essential change in the war–state relationship, as demonstrated by the Manhattan Project.

A significant difference between qualitative arms racing in the past and the experience of this phenomenon during the Cold War was that, in this case, it was freed from important constraints. In a hot war the pressure to get weapons and equipment into service as quickly as possible imposed a natural barrier to innovation. However, in the case of the Cold War this constraint was largely removed, and so a greater effort was focused on perfecting a weapon (Hartcup, 1993, 24). Indeed, in the context of nuclear deterrence, coming as close to perfection as possible reinforced the power of deterrence. The type of innovation that emerged in this setting was compared to Formula One racing where cost no longer matters and the pursuit of the last 20 per cent of performance frequently doubles the cost (Pugh, 1993). In sum, the geopolitical setting of the Cold War, combined with the opportunities and constraints created by technology, facilitated the rise of a relationship between national defence industries and the state, which was captured in the term the military–industrial complex (MIC). This condition reflected the dependence on the national armaments industry for the state to survive. At the same time, the state relied on its arms industry to produce the weapons technologies needed to deter and fight a war. This required the state to provide the right incentives for defence contractors to take on the development of technologically ambitious projects, which entailed a high level of risk and uncertainty. So great were these risks that defence production came to be seen as a unique form of economic activity representing a departure from the free market. This is why McNeill (1982) described the precursor to this period, which lay largely in the domain of the second industrial revolution, as command technology because of the role assigned to the state. Economists Peck and Scherer also claimed the weapons acquisition process represented a unique form of economic activity. As they explain:

The weapons acquisition process is characterized by a unique set of uncertainties which differentiates it from other economic activities. To be sure, uncertainty is a pervasive feature of all economic activity, and most of the uncertainties in weapons acquisition have their commercial counterparts. However, there is a uniqueness in both the magnitude and the diverse sources of uncertainties in weapons acquisition. (Peck and Scherer, 1962, 17)

An indication of the scale of these risks can be gauged to some extent from the difference in the average cost of research and development in defence, which was 20 per cent of sales, compared to the civil, and commercial sector, where in the 1960s, it was less than 3 per cent (Peck and Scherer, 1962, 45). Internal risks such as the possible technical failure of the project, as well as external risks, for example, a shift in government policy causing the programme to be cancelled, meant that state support for the defence industrial base was vital if it were to achieve security.

The key political and policy outcome of this venture demonstrated the vital role played by the state in allowing the talents of both private and public research and development to be unleashed by underwriting the costs incurred by defence contractors in conducting this process. This removed the risk of financial failure from the equation, which allowed more ambitious forms of research. Consequently, states came to play an essential part in a military version of Schumpeter's process of creative destruction, albeit in the realm of defence (1994, 82-3). The role of the state was vital because it provided the critical financial resources required to take embryonic technologies and develop them at a speed unlikely to be matched by the civil market. This facilitated a profound change in the relationship between the state and private industry and undermined the operation of the free market as governments opted to support defence contractors capable of conducting large and complex forms of research and development (R&D) (Chin, 2004, 43–69). The general assumption was that technological innovation was the product of a particular kind of industrial organization, what Galbraith (1967) called the giant corporation. He believed that only this kind of enterprise possessed the capital, research facilities and skilled workforce capable of bringing all these elements together in developing and manufacturing a new product. As he explained:

the more sophisticated the technology, the greater in general will be all the foregoing requirements. ... With very intricate technology, such as that associated with modern weapons and weaponry, there will be a quantum change in these requirements. This will be especially so if, as under peace time conditions, cost and time considerations are not decisive. (Galbraith, 1967, 28)

Although the relationship between the firm's size and the frequency and quality of innovation has been challenged, there is little doubt that this orthodoxy shaped the state's relationship with the defence market. Two important implications emerged from this assumption. The first and most immediate was that the pressures generated by the Cold War to innovate led state bureaucracies to channel their resources into large well, resourced defence enterprises, which led to industrial concentration: the erosion of the free market, and the emergence of a close and controversial relationship between the state and defence manufacturing. In the United States, defence manufacturing came to be seen increasingly as something unique in economic terms and was described as such. For example, Gorgol (1972) referred to the unique qualities of what he termed the Military Industrial Firm. Todd (1988) built on this concept to describe his variant, which was the military–industrial enterprise. This is why the term command technology is so significant; it changed the terms and conditions under which defence production operated and meant that it was not subject to the standard rules of demand and supply captured in the price mechanism and the free market. As a general rule, the state met the cost of R&D almost entirely. This was justified because it was not fair for the firm to carry the burden of risk and uncertainty that was an implicit part of weapons acquisition but, at the same time, it was also important that the state was not overcharged and the company did not make super profits on a contract designed to cover its costs and provide a reasonable profit. To ensure this happened, the state employed an army of accountants to audit company accounts. So, in a physical sense the war–state relationship remained alive and well.

The nature of this relationship and the implied position of privilege and power of the defence contractor within it led to the emergence of the idea of the MIC. First aired in 1961, President Dwight Eisenhower warned against the pernicious influence exerted by what he described as the creation of a MIC. This construct referred to the incestuous relationship between the military, defence industries, unions, local communities dependent on arms sales and politicians acting in concert as an interest group to convince the state to spend more on defence, even though the threat level did not always support such a diversion of resources.[1] This tied in with a wider debate on the increasing militarization of the United States during the Cold War. So, for example, according to one source, during the Cold War, the US Department of Defense (DoD) was the third largest planned economy in the world (Higgs, 1990, ix).

Harold Laswell noted the rising prominence of the military in peacetime in his thesis of *The Garrison State,* which described the potential militarization

---

[1]   Dwight D. Eisenhower, Farewell Speech to the Nation, 17 January 1961.

of the American polity (1997, 77–116). Samuel Huntington echoed this concern in his tome, *The Soldier and the State* (1985), which sought to reconcile how the United States could manage an immense military establishment in a time of peace without jeopardizing the sanctity of its democracy. These debates and themes waxed and waned as the Cold War progressed, but they persisted, and even in the 1980s, the notion of the MIC was still being aired. According to Darby and Sewall the evidence shows the United States became more militaristic during the Cold War. They refer to the national security bureaucracy established in 1947, which consumed an ever larger share of government spending. By 1964, defence research accounted for 17 percent of federal government spending. . They also point to the establishment of the DoD, the creation of an independent air force, the Central Intelligence Agency, the National Security Agency, the National Security Council, and non-defence agencies such as the National Aeronautics and Space Administration (NASA) (Darby and Sewall, 143, 2021).

However, the extent to which this represents a case of militarism can be questioned. The standard definition provided by Vagts states militarism:

> is a vast array of customs, interests, prestige, actions, and thought associated with arms and wars and yet transcending true military purposes. ... Its influence is unlimited in scope. It may permeate all society and become dominant overall industry and arts ... militarism displays qualities of caste, cult and authority. (Vagts, 1937, 12)

Most important is the extent to which the military can challenge by overt or covert means the values and norms of democratic governance. The existence of a MIC hinted at the possibility that the military and their allies across the wider economy and society did exert undue influence over democratic politics and policy-making. Although the power exerted by the idea of a MIC was challenged and debated, there is little doubt that this construct came into being due to the demands of the technological pressures generated by arms racing against a nuclear backdrop. As Berghahn (1984) noted in his study of militarism, the MIC represented a different form of militarism based on the existence of high-technology industries capable of constant innovation and the development of new generations of weapons. 'The salient feature of this military competition is that it is about very sophisticated weapons quickly becoming obsolete and constantly replaced by the next generation of even more sophisticated war materials' (1984, 88).

Most importantly, the resulting resource allocation to defence had a profound impact both economically and politically. This can be viewed on several levels. The most obvious is the sheer size of the US military establishment, which did not fall below 1.5 million military personnel during the Cold War. In the interwar period, it was 200,000 strong . At the next level is the resource allocation to defence, in the case of the UK, defence spending averaged over 6 per cent of GNP during the Cold War, in contrast, it spent below 2 per cent of GNP on defence. during the interwar period, even though it retained a global empire (Chin, 2004, 43).

Thus, although the shadow of total war and the mobilization of all resources via the state receded into the background post-1945, the tensions generated by the Cold War ensured the West remained on a semi-permanent war footing. The creation and sustainment of this infrastructure in defence production led to claims of a persistent form of militarism. The United States spent billions on expanding its scientific infrastructure. The Atomic Energy Commission, formed in 1946, took charge of the wartime test facilities and laboratories used to develop the atomic bomb. It also acted as the principal funding agent for the Lawrence Livermore National Laboratory. The Department of Defence was established in 1947 and sponsored what became a huge research programme. The National Science Foundation served as a conduit for civilian R&D. In response to the Soviet launch of Sputnik in 1957, the United States created NASA. Its sole task was to ensure that the United States won the space race. Much of this work was done in partnership with universities and the business sector. The state's role in exploiting defence research in the promotion of spin-off ensured this investment achieved a range of goals that extended beyond the realm's defence. The link between technological innovation and economic growth helped to create new industries, and generate employment and prosperity in general. Indeed, it is no exaggeration to say that defence research played a vital role in creating the computer and information technology revolution, which marked the shift from an industrial to an information-based society. So, for example, the development of computer hardware and software in the United States was primarily funded by the DoD, and it was not until the mid-1950s that significant private investment began to flow. In the case of IBM, government research contracts represented approximately 60 per cent of the firm's total R&D spending. Similar levels of government support were provided to Sperry and Burroughs. Between 1945 and 1984, the military invested US$81 billion (1972 dollars) in aerospace research. This compared with US$18 billion of investment from the private sector. Boeing's domination of the commercial airline market was attributed to the access it had to US government funding and its involvement in the design and manufacture of jet bombers such as the B 47 and B 52. The aero engine manufacturer Pratt and Whitney were also able to transfer DoD-sponsored research into the development of a range of commercially successful aero engines (Mowery and Rosenberg, 1989).

Most important here were the efforts made by the state to promote the diffusion of new technologies into the wider commercial sector. According to Stuart, 1993 the DoD facilitated conferences and financed engineering development at Bell, RCA and Pacific Semiconductor. It also underwrote the manufacturing facilities for Western Electric, General Electric, Raytheon and RCA and in addition played a critical role in establishing agreed manufacturing standards for these new technologies. For example, NASA instituted the Microelectronics Reliability Programme. In the 1950s, the US government also had a policy of supporting the entry of new firms into computers. A similar policy was employed in the domain of nuclear power. The Atomic Energy Commission provided subsidies and technology transfer to assist firms with the latent potential in entering this energy market sector. Finally, the civilian aviation authority made government-owned test facilities and test results available to industry (Stuart, 1993).

The belief that government could use defence research to enhance national security while at the same time promoting greater economic prosperity was a policy also followed by successive British governments during the Cold War. A case in point was the support given to the aerospace and electronics industry over many years. In the aerospace industry, it confronted an acute problem in that the UK national aerospace market was a fraction of the size of its US counterpart, and as a result, economies of scale were more challenging to achieve. US aerospace manufacturers also enjoyed a significant advantage in terms of the research funding they received from the government and they could divide their R&D costs over a larger number of units. Between 1955 and 1961, the average US production run for fighter aircraft was three times the size of the average of the UK – 530 versus 177 aircraft – which meant they also secured learning economies. As a result, American aerospace firms enjoyed a huge price advantage. To put this into perspective, in the 1960s, the US military aviation market was five times the size of NATO Europe. However, despite these structural disadvantages, the UK continued to support an industry that operated at a sub-optimal level of efficiency. To address this deficit, the UK government relied increasingly on collaborative ventures with other European partners that sought to combine national orders and create a level of productivity and efficiency that might allow it to compete with US defence manufacturers in the global aerospace market. In following this path, it was estimated that defence collaboration would create a market three times as large as the UK national market. An important driver of this policy was a strong desire to preserve a national aerospace industry. This reflected a mixture of political motivations, but of importance, in this case, was the significance of aerospace as one of the

UK's critical high-technology sectors. It was noted that the problems facing UK aerospace vis-à-vis the giant size of the US aerospace market were shared with other UK areas of high-technology and that, as a UK government report observed: 'For this reason we cannot afford to admit defeat in this particular industry while hoping that solutions will be found in others. Rather we must seek a solution here which may have a more general application' (Elstub, 1969).

An important difference between the UK and the United States was the extent to which the state became enmeshed in reorganizing the British aerospace industry and other high-technology sectors to create an industrial structure better able to carry out complex R&D. This meant promoting or directing mergers and acquisitions in the defence market, eventually creating national monopolies in airframes, aeroengines and defence electronics. So, for example, although in the 1980s, the UK Ministry of Defence had over 8,000 suppliers on its approved suppliers' list, 20 per cent of suppliers won 80 per cent of contracts (Bittleston, 1990, 15). It was estimated that a large slice of this work went to only five contractors: Ferranti, GEC, BAe, VSEL and Plessey (Chin, 2004). The British also briefly flirted with attempting to elevate this process on a pan-European scale in an effort to make the European defence industrial base more competitive and, in line with the Single European Act, which sought to promote free trade across the European Economic Community, create a high-technology sector capable of competing with the American Emerging Technology Initiative and the Strategic Defence Initiative in 1982 and 1983 (Vredeling, 1987).

Why, by the mid-1980s, had state-sponsored innovation in defence ceased to be the vanguard of the technological revolution and what were the implications of this for the war–state relationship? This was particularly true in computers, semiconductors and software development, where traditional defence contractors struggled to get to grips with the rapid change taking place in this product area.

Defence's loss of its technological lead can in part be attributed to the success of promoting spin-off, and the private sector exploited the scientific research it had acquired to develop products that have shaped the material world in which we live. This includes GPS-enabled technologies, airbags, lithium batteries, touchscreens, and voice recognition. All of these capabilities were initially the products of government investment. However, it is also clear that the nature of defence research meant that it focused on bespoke capabilities that were important to defence but had little relevance to the civil and commercial sectors, for example, munitions and certain materials technologies used to improve the protection of armoured vehicles. At the same time, it is also clear that civilian technology firms also adopted a less ambitious technological horizon. A peculiar feature of the weapons acquisition process is that it takes so long to develop and deploy high-tech weaponry. For example, the European Typhoon fighter was first demonstrated as a technology in 1980 but did not enter service with the RAF until 2004. Given these long time horizons, it is inevitable that defence needs to think in terms of technological leaps to ensure the weapon is not obsolete before it is deployed on the frontline.

In contrast, the commercial sector has a much less ambitious timeline and seeks to harvest all low-hanging fruit in terms of what is available in terms of technology. As such, companies tended towards an incremental development model, one based on the exploitation of existing technology but made smaller, lighter, faster or more energy efficient. It is also claimed that smaller, nimbler research organizations increasingly outperformed the military–industrial firm dominated by an obsession with size. These were startups funded by venture capitalists. The idea of venture capitalism emerged in the 1970s, but it was not until the 1990s that this form of investment gained traction. It is claimed that large companies spend less on internal R&D and more on corporate development, or acquiring smaller venture-backed companies that are developing novel and exciting technologies (Darby and Sewell, 2021, 144). In addition, government spending on R&D declined as a percentage of GNP, which happened at the same time as private sector research funding started to increase.

Conclusion: Why did the pursuit of technological excellence persist after the Cold War?

The dramatic end of the Cold War in 1989 and the collapse of the Soviet Union in 1991 transformed world politics and created a world system dominated by the United States, albeit briefly. Within this global Pax Americana, the threat of a major war disappeared, but the condition of war persisted, albeit on a smaller scale, as a series of civil wars erupted as a direct consequence to the end of the Cold War. Europe was not immune to this trend, and it experienced its first major conflict since the end of the Second World War with the collapse of Yugoslavia and its descent into civil war in 1992. This conflict became the paradigm for what were described by contemporary observers to be 'new wars' – a term that was strongly contested (Fleming, 2009). These conflicts were associated with the impact of globalization which exposed their economies to the effects of global free trade which induced significant financial and political stress. Compounding this fiscal stress was the decline in the level of support offered by the superpowers during the Cold War. In the new world order, these regimes were now left to fend for themselves. This produced internal instability, and unscrupulous politicians sought to take advantage of this moment of crisis by mobilizing domestic support through the exploitation of identity politics, which manifested as ethnic conflict and civil war. An important component of the new wars thesis was the role played international actors in this type of conflict (Kaldor, 1998). In looking at the international response it is striking that, in the case of Yugoslavia, 365,000 people lost their lives and over two million were either internally displaced or became refugees.

However, this humanitarian crisis produced a muted and erratic response on the part of the European governments and the United Nations. The deployment of a UN peacekeeping force did little to stop the violence and it became clear that national governments were not willing to adopt a more robust policy towards Serbia because, while willing to deploy peacekeepers, they did not believe this crisis justified becoming embroiled in a war that mattered little to them in material terms. This political miasma was present in other UN peacekeeping missions in Somalia (1992–94), Rwanda (1994) and initially in Kosovo in 1999. Governments felt compelled to become involved in humanitarian interventions but these wars were described as 'wars of choice', meaning there were no vital national interests at stake for the governments whose forces were intervening to preserve the peace. Peacekeeping in the post-Cold War era revealed many of the same societal reservations over the costs of war which arose in the wars of decolonization discussed earlier.

This disconnect between the need to do something and the reluctance to do so was reconciled through increased reliance on technology which had been validated by the outcome of the First Gulf War in 1991. All through the Cold War, analysts questioned the wisdom of spending so much on defence research for technologies which cost a huge amount but remained unproved in the reality of war. At the same time, both US and Soviet military analysts were exploring how the computer and communications revolution would impact the conduct of war. The Americans distilled their thoughts in their military doctrine known as AirLand Battle. In the case of the Russians, they referred to a military–technical revolution. The campaign to liberate Kuwait provided the evidence and reassurance that quality and not quantity counted in battle as the US-led coalition routed the Iraqi military in a war in which the headlines focused on the technological superiority of the US military. Not surprisingly, the Americans sought to capitalize on their technological lead and so, lacking a clearly defined enemy against which to build, initiated what became known as the revolution in military affairs (RMA). Maintaining their technological lead became the default setting of US military planners, as expressed in Joint Vision 2010 and 2020, which were published in 1996 and 2000 and their allies did their best to keep up fearing that if their systems were unable to communicate with US Command Control, computers and communications, intelligence, surveillance, and targeting and acquisition systems they would cease to have any value. The fruits of this technological revolution provided at least a partial solution to the political challenges posed by humanitarian intervention by allowing Western forces to engage the enemy at a distance that ensured their military personnel were not in harm's way. New terms were employed to describe this latest technologically based military revolution. McInnes (2013) described this as spectator sports warfare. Ignatieff (2001) called it Virtual War. Shaw (2005, 38–58) argued that it was the new 'Western way of war', which he linked to economic and societal conditions within Western states captured in his idea of the 'post-military society'. The best illustration of this form of war was the air campaign launched to stop the Serb government from ethnically cleansing the local Albanian population from the province of Kosovo in 1999. In this conflict, NATO governments refused to sanction an invasion to stop Serbia from ejecting the local Albanian population and so was forced to resort to the use of airpower to achieve its objective. A further constraint imposed on the air campaign was the need to ensure no NATO pilots were lost to Serb air defences, and so bombing was conducted at 15,000 feet. Although it took it 78 days to force the Serbs to comply this was declared a victory and the first ever war won by airpower alone. Most important, NATO did not lose a single pilot in this operation. The reliance on technology also provided a neat solution to the challenge of how to wage a global war on terror. The war against the Taliban in 2001 demonstrated how through the use of real-time surveillance technology both on the ground and in the air it was possible to deliver precision-guided missiles and bombs to their targets in real-time. This campaign was seen, at the time, to be a revolutionary moment in war. With fewer than 300 CIA and Special Forces on the ground in Afghanistan the United States was able to bring about the defeat of the Taliban in less than three months. This campaign seemed to show that the RMA could prevail even in an asymmetric war. In the aftermath of this war a great deal of faith was placed in technology to allow the United States and its allies to fight a range of different opponents. One need only look at the technology driven doctrines that came to the fore, for example networked enabled capability and effects based operations (Mathis, 2008). Although the assumptions underlying these doctrines was sorely tested as the war on terror took on a darker hue, first in Iraq and then Afghanistan, innovation continued. Of particular significance was the expansion of automation witnessed in the dramatic increase in the number of remotely piloted vehicles (RPVs) and ground-based robots used to prosecute this war. Singer (2009) referred to a robotics revolution in the conduct of war. By 2012 the US military went from having only a small number of UAVs in 2001 to over 12,000. It is also at about this time that the process of automating war took on a new significance as a range of new technologies appeared to be on the cusp of becoming a tangible reality in the world of defence. The next chapter explores how the military plan to push towards the next military revolution.

5

The Western Military Vision of Future War

The military is frequently accused of using the tactics of the last war to fight the next, a tendency that has sometimes led to disaster, for example, the Fall of France in 1940 or the more general catastrophe that befell all nations in the opening phases of the First World War (Strachan, 2005, 156–87). In both instances, failure was believed to be a consequence of military doctrines that bore little relationship to the prevailing realities of contemporary war. The modern-day military as an organization knows and understands the need to control this behavioural trait and has made a concerted effort to avoid repeating past mistakes. However, one thing that has not changed in the Western military establishment is a continued fascination with technology as a force multiplier when thinking about future conflict. This perceived obsession is strange because it feeds into a broader debate that focuses on the role played by technology in facilitating the defeat of Western militaries in a series of what might be loosely described as irregular wars since 1945 (van Creveld, 1991; Lyall and Wilson, 2008). However, despite the experience gained from two decades of war waged by Western states across the Middle East and Central Asia, the one salutary lesson learned from the war on terror, including the war against Islamic State, is that, even in irregular warfare, technology has its uses. However, it is no substitute for a strategy based on an understanding of the human domain of conflict, which is manifest in the political, ideological and economic strands of power. The challenge, then, is to ensure technology does not become a substitute for other more appropriate forms of power, which implies a more nuanced use of this capability. In a sense, the debate between technology and the operational context in which it is employed – a conventional interstate or an unconventional intra-state war has become increasingly irrelevant as recent conflicts have become hybrid in nature. This blending of conventional and unconventional is itself seen in part as a consequence of new technologies enabling non-state actors to acquire state-like military capabilities via the adaptation of commercial off-the-shelf products and services (Niam, 2014, 107–21). Equally important, however, is how states, such as China, Russia and Iran, have embraced and incorporated elements of irregular warfare into their strategies, and how these have also been blended with their exploitation of technology (Kilcullen, 2020). All of this suggests that the debates between regular and irregular war have become a distraction. Russia's invasion of Ukraine in 2022 highlights this point. On paper this war looks and feels anachronistic – because it is being fought between two states for the most traditional of reasons, the conquest of territory, using the massed concentration of 200,000 troops to achieve its objective (Fazal, 2022). In sum, it is a very traditional Clausewitzian war. It is also a war that features elements of pre-modern war in the form of mercenaries both from within Russia's Wagner Group and Syrian volunteers sent by the Assad government as payback for Russian support in Syria's civil war. The salience of technology in shaping this conflict has been pronounced. In the first phase of the war, which focused on the rapid conquest of Kyiv and the overthrow of the Ukraine government, the Russian offensive stalled because of the exploitation of innovative technologies by the Ukrainians in the form of drones, guided surface to air and anti-tank missiles and even drawing on Elon Musk's Starlink satellite system to ensure Ukraine has access to the internet. In sum, whatever perspective is taken, war in the 21st century is being defined by technology both on and off the battlefield. In this chapter, I want to look specifically at how the Western military views the challenges and opportunities posed by recent conflicts and how their observations of recent wars have shaped their thinking about the role of technology and the state in possible future wars. As in previous chapters, my central argument remains that contrary to popular belief, technology continues to define and shape the war–state relationship and that new technologies may have precisely the opposite effect in expanding the interaction between them – a theme strongly implied in the latest iteration of military doctrine captured in the construct of multi-domain integration (MDI), which is the principal focus of this chapter. Military doctrine is essentially the military's best guess of what a future war will look like. In following this train of thought it is important to take on board Colin Gray's observation:

> Future warfare can be approached in the light of the vital distinction drawn by Clausewitz between war's 'grammar' and its policy logic. Both avenues must be travelled here. Future warfare, viewed as grammar, requires us to probe the probable and possible developments in military science concerning how war could be waged. From the perspective of policy logic, we need to explore official motivations to fight. (Gray, 2005, 39)

This chapter focuses on the military's conception of the policy logic of war followed by an exploration of what this means in terms of war's grammar.

It is important to do it in this order because the policy logic of war plays a vital role in highlighting the military judgement concerning the future context in which force will be employed and how it will be used. As such, I set out the political context in terms of why and by whom war will be fought and the role played by technology in this future vision. Within the scope of this chapter, we need to think about how reactionary or progressive the Western military conception of future war is, how it seeks to employ new and emerging technologies, and the implications of this in terms of the war–state relationship.

Most important are the conditions that challenge a vision which looks and feels very traditional. In essence, how does the US military continue to justify the centrality of battle in an era of diffuse threats? In scrutinizing war's grammar, the intention is to provide more than a mere speculative description of emerging technologies within the future battlespace but to demonstrate how precisely the military intends to use these tools in a future war. Analysis at this level is critical because it feeds into a broader debate about the culture of Western militaries and, in particular, the claims made about the emphasis they place on war's grammar at the expense of its policy logic. At first glance, the UK and US visions of future war reinforce this stereotype. This then implies a further confirmation that doctrine is an intellectual product largely concerned with the internal logic, culture and interests of the military and is only coincidentally about addressing the external threat, a debate which has a long history in the field of innovation and adaptation (Posen, 1984). This is an issue explored in this and the next chapter.

Revival of great power competition and qualitative arms racing

Contrary to popular perception, contemporary military organizations have demonstrated an increased awareness of a range of security challenges, many of which lie outside the traditional warfighting domain (DCDC, 2017). At the same time, military treatises and doctrines have also revealed an increased recognition that the division between hard and soft security threats in the form of migration, climate change and pandemics are not separated but often interlinked with warfighting. For example, it is claimed that climate change caused and exacerbates conflict in places such as Syria and Darfur. In the case of Syria, the civil war, which has been attributed in part to drought, led to economic distress in the rural areas and a shift of the population to the cities, which in turn provided the conditions that fed the fire of revolution in 2011. This war then created a power vacuum in the east of the country, which the Islamic State and other violent non-state actors exploited in their efforts to access its resources. This, in turn, facilitated external intervention as the United States and its allies sought to destroy Islamic State in Syria and

Iraq. As important, it also allowed both Russia and indeed Iran to extend their influence and frustrate Western policy ambitions to overthrow the Assad regime. In the case of Russia, its activities here have been connected back to its efforts to secure its position in Ukraine by forcing NATO to deploy scarce resources to its southern flank in the Mediterranean. As such, the traditional dividing line between internal and external conflict has become blurred, and so too has the division between conventional and unconventional war as states use both state and non-state actors and have resorted to a variety of means, regular and irregular, to achieve their aims.

Equally important is the role played by technological advances and the diffusion of power that blurs the boundaries of war in terms of internal and external threats. A cursory glance at the US National Defense Strategy published in 2018 demonstrates a broad awareness of the multiple challenges facing the United States and its allies. For example, it recognizes that the civil and commercial sector now creates so many technological innovations that capabilities beyond the means of small states and non-state actors are now within their grasp. This means smaller states may acquire the capability to challenge America's current military superiority. It also means state and non-state actors will employ technology asymmetrically to bypass rather than challenge US power. Most importantly, there is a broader understanding that 'new commercial technologies will change society and, ultimately, the character of war' (US Department of Defense, 2018, 3).

However, recognizing and understanding the complexity of this mélange of emerging security threats has not resulted in a dramatic shift in the focus of military thinking about future war. If anything, it feels like it has regressed back to the era of the Cold War with its emphasis on state-based threats, on deterrence and proxy wars. As in the past, current military thinking on future war appears to be fixed on the high end of the conflict spectrum: major interstate conflict between the leading economic and/or military powers. Is this merely evidence of reactionary thinking, in essence, a manifestation of the desire to use the tactics of the last war to fight the next? From the perspective of the military the answer is no. However, it is clear the West feels threatened by Russia and China. Authoritarian governments represent a menace to the norms and values of the West and the world order created by the United States at the end of the Second World War. The democratic wave that followed the end of the Cold War is now receding. Moreover, technology is being exploited by China and Russia to create something approaching a surveillance state, and the systems and ideas crafted to achieve control are being exported to other non-democratic regimes. Authoritarian regimes also enjoy certain political advantages because the centralization of government makes it easier for their leaders to control the levers of power, potentially allowing for the creation of a more effective grand strategy. This is not mere rhetoric articulated to justify certain forms of military power. The view that the West is locked in a conflict to save democracy is one echoed by a range of sources outside of the military (Kagan, 2008; Allison, 2018).

In the case of China, its rise has led it to become a threat to the dominant position of the United States in the Asia Pacific region. China's GDP, which is a basic measure of power between states, was only a fifth of that of the United States in 1991, but in 2020, was a fifth larger, when measured by using a purchasing parity index (Allison, 2018, 33). There is a consensus that China is using its newfound economic might to coerce Western allies in the Asia Pacific region and has become a global competitor to the United States. The challenge posed by China is not just geopolitical but extends into the technological domain. In 2007, Will Hutton wrote a book that highlighted the many internal economic problems confronting China, but what was most interesting was his observation that China had failed to evolve as a centre of technological innovation. At that time, he concluded that, unlike Japan which had managed to transition from the metal bashing end of the manufacturing spectrum to become a fundamental driver of innovation and value-added manufacturing, China was expected to fail in making this transition (Hutton, 2007). Viewed from the present perspective, one of the most impressive aspects of China's rise has been its success in innovating in high-tech industries once dominated by the West.

In global terms, nine of the world's largest technology companies today are Chinese. In the field of AI, Google, which remains the world leader in this field, is on the verge of being caught by Chinese companies, and their global dominance seems assured because they can tap into the data of 1.3 billion people, which will dramatically increase the ability of Chinese AIs to learn and grow in power. Indeed, it is predicted that China will overtake the United States in this field as the state pours money into this research domain (Lee, 2018). China's economic miracle has provided the means for it to project power on a global scale. For example, it is spending US$1.3 trillion on infrastructure for the Belt and Road Initiative, which will link most of Eurasia to China economically. In response, the United States offered a more modest package of measures intended to counter China's expansion: US$133 million was pledged in new investments in the Indo Pacific region. China also made a sustained effort to invest in improving its military power. Twenty-five years ago, Chinese defence expenditure was less than 5 per cent of the United States'. Today it spends about a third, and it is feared that China will soon reach parity with the United States (Allison, 2018, 34). Most importantly, China's spending is concentrated mainly in its region, whereas US military spending is intended to provide global coverage.

In contrast, the threat posed by Russia is not about problems relating to power transition within the global system but precisely the opposite. Russia is a power plagued by the shadow of its former glory, its current diminished status and the belief that it is threatened by the West's creeping expansion eastwards, symbolized by the encroachment in former Soviet space by the European Union and, more importantly, NATO. To the Russians, the physical and spiritual drift of former Soviet republics toward Western liberal institutionalism is perceived as a threat to the internal stability of the Russian state. In addition, there is also a sense of anger because it feels betrayed by Western failure to adhere to an agreement after the fall of the Berlin Wall, which stipulated that the West would respect Russia's traditional sphere of influence in its near abroad. In response, the Putin government and the Russian military establishment concluded that Russia is threatened not merely by the presence of NATO but even more so by the allure of the West as expressed in democratic governance and the prosperity it offers. These fears were given substance with the onset of the Arab Spring in 2011. This spontaneous uprising across the region and the recent memory of how the Soviet Union's collapse was precipitated by internal division compounded the sense that it faces an existential crisis that is political, social and cultural, and physical (Fridman, 2018).

Paradoxically, the West also feels threatened by the international order that is now emerging. Two decades of protracted and financially ruinous war against al Qaeda and its allies harmed the West's strategic position and reputation. Compounding this increasing anxiety is the fear that this broad coalition is declining in economic terms. The 2008 financial crash tarnished the reputation of Western economic competence, and its insistence that the free market is the best and only model to pursue by developing states seeking to modernize has weakened its appeal and undermined its soft power. Technology, in particular the rise of social media, has also provided a megaphone that has allowed the discontent caused by a decade of stagnation in living standards in the West to surface in a way that is challenging traditional forms of governance. The cunning use of media, first by al Qaeda in the Arabian Peninsula and then Islamic State, also demonstrated how the West's enemies could exploit social media to promote their cause and provoke internal violence within Western metropolises. Both Russia and China have sought to exploit this fissure within the West.

The military focus on 'great power competition' as opposed to the threat posed by terrorism, pandemics or climate change is also justified in terms of it being the most immediate threat facing the sanctity of Western states and, as important, a task that falls within the domain of the military in terms of its role as guardian of the state. Russia and China have invested heavily in defence over the last two decades. China is currently the second highest spender on defence in the world. In March 2023, the government announced a defence budget of RMB 1.5 trillion ($224 billion) – which represents 7.2 percent increase over the previous year. Because of differences in how governments define and measure military expenditures there are always variations. In the case of China, it is believed the official data presents an underestimate and SIPRI calculate that a more realistic figure is $285 billion.

What is also striking is how quickly China's defence budget has grown, it has doubled in cash terms since 2013. Approximately 40 percent of this budget is invested in the procurement of equipment and new technologies (CSIS, 2023). The US defence budget still dwarfs that of China and stands at $842 billion in 2023. However, as the Economist demonstrated, when using a purchasing parity index, the gap between China and the United States closes when you look at the buying power of their budgets.

So, for example, the entry costs for a soldier in the US Army are 16 times higher than in China. Using this calculation method allows us to see that Chinese defence spending was more than double the sum estimated by SIPRI in 2021 and stood at a whopping US$580 billion, which is a lot closer to US expenditure on defence (*The Economist*, 1 May 2021).

In the case of Russia, before 2021 its defence spending increased 21 years consecutively and stood at approximately $63 billllion in 2022. On average, between 40–50 per cent of the defence budget was spent on equipment, support and maintenance and modernization of its forces. This fiscal activity was seen a cause for concern by NATO and these fears were compounded because, as in China, there was a belief that the purchasing power of the local Ruble meant that its official figures on defence spending were misleading. Again using a purchasing parity index indicated the size of the Russian defence budget was closer to $150-180 billion (CRS, 2020; Kofman and Connelly, 2023). One reason why Russia's material superiority failed to overwhelm Ukraine was because of the indirect support provided by NATO nations. As such, Ukraine has become the West's proxy war which aims to contain perceived Russian aggression. Faced by the prospect of failure and diminution of its influence in former Soviet space, the Russian government is determined to fight on and the war planned to increase defence spending by 40 percent in 2023 in the hope that it 'can snatch victory from the jaws of defeat' (Luzin, 2022).

The new world order and its impact on war, technology and the state

Defence research and the role of the state

It is not just that states maintain high levels of defence spending; as important is what they are spending this money on. As in the Cold War, a considerable sum is being invested in trying to achieve a step-change in military capability to ensure forces have a decisive edge over their opponents – in essence, a return to qualitative arms racing. For example, in 2017, the United States spent over US$55 billion in supporting defence research. In the case of the UK, this was a more modest US$1.5 billion (CRS, 2020).

The reasons for this are straightforward. First, as has been said, there is a solid inbuilt bias in the military to search for a technological silver bullet that promises to transform war and provide a decisive advantage in battle. It is also evident that the application of advanced technologies addresses the real political constraints that limit the use of force in the Western world and compensates for the decline in the size of militaries as fiscal prudence bites into defence budgets. The second is how states are currently seeking to employ and exploit defence research within the broader context of their national technology and economic strategies. China has a long history of using the state's power to promote a more comprehensive technological and economic agenda (Tai Ming, 2022). Historically the United States continued to support a state-led solid defence research programme. Indeed, the Reagan presidency was accused of using military spending to promote a form of military Keynesianism. This included exceptional help in developing dual-use technologies such as semiconductors. In the case of the UK, the current government's thinking on defence research and its direct value to the wider economy has changed. The quotation taken from the UK's latest defence review, *The Integrated Review*, makes this aspiration clear:

> Our first goal is to grow the UK's science and technology power in pursuit of strategic advantage. Achieving this objective requires a whole of UK effort, in which the government's primary role is to create the enabling environment for a thriving science and technology (S&T) ecosystem of scientists, researchers, investors and innovators, across academia, the private sector, regulators and standards bodies, working alongside the manufacturing base to take innovations through to markets. It also requires strategic choices and decisions by the government, both on S&T priorities and on how we use our national S&T capability to support wider policy goals. (HM Government, 2021, 35)

The importance of technology will increase because of the following factors. First, as highlighted in the introduction, the pace of technological change is increasing rather than slowing. Developments in AI and quantum computing provide new means to accelerate scientific and technological innovation. Second, S&T will become an essential power metric, conferring economic, political and military advantages. The tech superpowers are investing in maintaining their lead. At the same time, many more countries can compete in S&T, while extensive technology companies can grow more powerful by absorbing innovations produced by small companies. Competition is therefore intensifying, shaped by multi-national firms with the backing of states, some of which take a whole of economy approach to ensure dominance in critical areas.

> Maintaining a competitive edge will rely on preeminence in and access to technology – as well as access to the human and natural resources

needed to harness it – and the ability to protect intellectual property. As the volume of data grows exponentially, the ability to generate and use it to drive innovation will be a crucial enabler of strategic advantage through S&T. (HM Government, 2021, 24)

Technological innovation, therefore, promises to address a wide range of objectives that extend beyond the narrow realm of defence, but it is in this area that increasing anxiety is being aired over the decline in the West's technological lead. In 2015, Mark Urban warned that the West was losing its technological edge in relation to the Russians and ominously sought to compare the age of austerity in Western democratic states post-2008 financial crisis to the interwar period within Europe. His tome essentially repeated the old mantra. Kilcullen (2020) observed that the Russians and Chinese were not only catching up technologically but also developing military doctrines based on their observation of the Western way of war that unfolded in the Gulf in 1991. In response, they crafted their technological asymmetries that sought to neutralize the principal elements of Western military superiority. This was particularly true in missile defence, which limits the ability of the United States to conduct offensive operations. In addition, Russia and China have made huge advances in cyber and the Chinese have ventured into space, sending RPV space vehicles to the moon and, in 2021, to Mars. The last is particularly worrying given the heavy reliance placed on satellite communications by the US military and their allies, and they fear a future war will also incorporate attacks against this vital infrastructure. In sum, the great powers are relying on technology to enhance the deterrence of their forces. This in itself is not new, and history is replete with examples of such action–reaction cycles (Brodie and Brodie, 1973). However, this has generated a dynamic but unstable technologically driven arms race as each seeks to counter the advantages of its adversary through technological innovation, which feeds back into pushing for more significant innovation. The efforts of the Russians and Chinese to deconstruct US power projection and its ability to refight Gulf War 1 and 2, followed by the current US effort to counter this via its new technologies, provide a classic illustration of this phenomenon.

A vital component of this current arms race focuses on the exploitation of AI. Putin declared that whoever became the leader in the development of AI would become the next ruler of the world. It is claimed China wants to achieve a position of global dominance via this domain by 2030. China and the United States are currently at the head of a pack of states, including Israel, India, Russia, Germany, the UK and France. For this reason, the current AI arms race differs from arms racing during the Cold War where competition was mainly between two states, which reduced the scope for error and misunderstanding. Moreover, both sides reached a point where the pursuit of military advantage was so ephemeral and destabilizing that acceptance of a technological status quo became the best option. In the case of AI, no such logical end state exists (Kissinger, 2018).

According to a report published in 2020:

> technology development and acquisition – especially in AI – is a critical part of this expanding competition between the United States and China. Indeed, this competition extends into semiconductors, 5G networks, robotics, quantum computing and neuroscience. These technologies are perceived to be central to facilitating an economic and social transformation, but also providing the means to control it. Most important is the role of AI and the other technologies listed in facilitating the modernization of China's military capability. (Nurkin and Rodriguez, 2019, 10)

In seeking to restore Western technological superiority, the United States drew on its own recent history to explore how best to exploit the vast technology pool of expertise that exists within its economy. In doing so, it identified what were described as two previous offset strategies that had been employed by the United States to establish a qualitative superiority during the Cold War. The first of these offsets sought to balance the numerical superiority of Warsaw Pact forces facing NATO by relying on a more powerful generation of nuclear weapons to deter aggressive action articulated in what was called the New Look. The second sought to use the information and technology revolution in the 1970s to counter not only the numerical superiority of Warsaw Pact forces but also the increasing quality of its equipment and armaments (Mearsheimer, 2017). The third offset sought to use AI, cyber, robotics and machine learning to allow the defeat of both Russian and Chinese defences in the event of a war (Ellman, 2017). It is claimed that the third offset was enshrined in the United States National Defense Strategy 2018. We can also see strong echoes of this in the US military's efforts to develop an appropriate doctrine and suite of capabilities to fight a future war based on the logic and assumptions of this latest concept of operations known as MDI.

Another essential difference between technological innovation during the Cold War and today concerns the state's and industry's very different roles. During the early days of the Cold War, when the infrastructure for command technology was firmly established in the United States. As such, the state played a critical role in facilitating the acceleration of scientific research in general and specifically in defence during the Cold War. The private sector did, however, benefit as it capitalized on these technological innovations and applied them to the broader commercial sector. A significant evolution in the US state industrial relationship in defence research was the increasingly important role played by the private sector, initially in the form of large corporations. However, more recently, it has been in the form of smaller, privately funded startups. These companies have attracted significant funding from venture capitalists and became increasingly important in the defence sector in the 1990s. According to Darby and Sewell (2021), venture capitalism has taken on more of the burden of research as it shifted from government-owned research facilities to large private corporations and, in this sense, such companies have a vital role to play in the development of defence technology.

However, the need to make a quick return on an investment in a new technology inevitably ensures that the time horizon of projects funded by venture capital is no more than ten years. As a result, the private sector has not been interested in strategically essential technologies such as microelectronics, a product sector where profits emerge after decades rather than years of investment. The net result is that funding for defence research does not always align with national security priorities (Darby and Sewell, 2021, 144). The most apparent evidence of how the balance between state and industry is changing can be illustrated in the case of NASA and its partnership with Elon Musk's company Space X. Between 1969 and 1972, NASA successfully landed six men on the moon. During the Trump Presidency, it was decided to revive this mission as part of a longer-term goal to send a piloted mission to Mars. Various reasons were cited for undertaking this mission, but competition with China and the need to develop technologies that will allow the United States to ensure it can protect its estimated 2,500 satellites in space in time of war play an essential part in this project. Of importance here is that NASA is relying on Elon Musk's Space X to develop a lunar lander at a cost of US$2.9 billion. Space X has already provided the means to deliver US astronauts to the International Space Station.

China is also making a sustained effort to develop advanced technologies in defence. Like the United States, its principal areas of interest are AI, quantum computing, and deep space, deep sea and polar exploration. Spending on basic research increased by 10.6 per cent and was set to grow by 7 per cent over five years from 2018-23. A key aspect of its latest five-year plan is the focus on the modernization of its weapons and equipment as well as the promotion of an independent defence technology base. The Chinese political and economic system has played a vital role in facilitating innovation. Interesting from the perspective of this book is the important role played by the Chinese state in mobilizing and directing the:

> resources of its science and technology community in a way that is not easily replicated in the United States. The central direction and control have implications for collaboration between civilian and military enterprises, for data collection, storage, and use, and

for the acquisition and transfer of technologies from the civilian sector to the military through military-civilian fusion. (Nurkin and Rodriguez, 2019, 11)

The importance of technology in the eyes of China cannot be overstated. Over the past two decades it has invested heavily in R&D. Its share of global technology spending has grown from 5 per cent in 2000 to over 23 per cent in 2020. Based on current trends, China was expected to overtake the United States in spending on scientific research by 2025. The Chinese approach differs from the United States in that it fosters the development of a wide range of technologies, not just those that are specific to the security domain. Although much of this work is conducted by private companies, the state provides the lion's share of the funding through grants and government-backed loans. It is claimed the Chinese are building whole new cities dedicated to the goal of innovation. According to Darby and Sewall Huawei, which is the world leader in the provision of 5G networks, has received an estimated US$75 billion in tax breaks, loans, grants and discounts. It has also received generous support for its work on the Belt and Road Initiative. Because of this support, Huawei offers cheaper 5G technologies than any of its competitors:

> Washington has monitored China's technological progress through a military lens, worrying about how it contributes to Chinese defence capabilities, But the challenge is much broader. China's push for technological supremacy is not simply aimed at gaining battlefield advantage; Beijing is changing the battlefield itself. Although commercial technologies such as 5G, AI, quantum computing, and biotechnology will undoubtedly have military applications, China envisions a world of great power competition in which no shots need be fired. Technological supremacy promises the ability to dominate the civilian infrastructure on which others depend, providing enormous influence. (Darby and Sewall, 2021, 148)

In contrast, the United States tends to equate security with traditional defence capabilities, so vital technologies are not being developed. For example, microelectronics is a critical component in defence and the wider commercial sector, and AI's advent has amplified its importance. However, investment in this vital capability has fallen drastically, and neither the public nor the private sector are providing sufficient investment in this sector. An important reason why this has happened is because of the reliance placed on the private sector, which is put off by the large-scale capital investment required and long time horizons before it makes a profit on its financing (Darby and Sewell, 2021, 149).

A note of increasing concern in the United States is that its version of command technology is inferior to that employed by the Chinese state, which has played a leading role in China's wider economic development and its transition from a country that imitates to one that innovates in technology. As has been pointed out, such a strategy produced considerable waste and inefficiency, and although it looks like its American counterpart in that it is a mixed bag of state and private sector investment, there is an important difference between the US and Chinese models. This lies in the private–public relationship, as one observer commented. In the US high-tech sector, there is a degree of distrust and hostility towards involvement in military research. The best illustration of this is the petition signed by 3,000 Google engineers who opposed the firm's involvement in Project Maven. No such reservations appear to afflict China's state industrial relationship in areas like AI (Cox, 2021).

The new grammar of war

A return to great power competition contains little that is new. Similarly, qualitative arms racing, which focuses on achieving dramatic improvements in weaponry, has also been a part of the fabric of war since at least the mid-19th century and reached its apotheosis during the Cold War. What is different today lies partly in the speed and breadth of change. It is not just one technology that is changing, and it feels as though every aspect of our lives is increasingly shaped by technological change. However, the most important feature of this latest spasm of innovation lies in the way it is causing the diffusion of power within and across societies and the creation of new vulnerabilities within nation states and this has led to a bifurcated response from the military in terms of how to address these new challenges. This diffusion of power is most evident in the increased accessibility innovation is generating in terms of cost. In essence, the correlation between cost and performance is no longer as fixed as it once was, and many capable systems are falling in price. This applies to both the commercial and defence sectors. This has happened due to dual-use technologies, which allow cheap mass-produced components to be used in defence production to enhance or replace more complex and expensive weapons systems. This means that the nation state's monopoly on the use of force is being eroded, and there is a real danger that Schumpeter's process of creative destruction could result in a new idea, product or process that challenges the world military order and the dominance of the West (Schulke, 2022).

Most important dual-use technologies allow states and non-state actors to weaponize even the most benign capabilities. Much of this can be seen in within the domain of political and information warfare, which exploits social media to expose internal social and political contradictions within

Western states. We have seen Russian intervention in the US presidential elections in 2016 (Rid, 2021) and accusations that the victor of that election, Donald Trump, owed his success to Russian state-backed groups creating fake news on the internet. Similar accusations were made regarding the outcome of the UK referendum on whether to leave or remain in the European Union in 2016. Again, the outcome of that vote seemed to favour Russian interests because it divided Europe and weakened the European Union. Islamic State's strategy for defeating the West also exploited social media and their campaign resulted in an unprecedented number of foreign fighters flocking to Iraq and Syria to fight for their cause; they recruited more fighters in two years than the Mujahedeen in Afghanistan did in over a decade of war against the Soviet Union (Stern and Berger, 2015). Fears of the weaponization of the internet continue to be a cause of concern today. Interestingly, the UK's former Chief of the Defence Staff, General Sir Nick Carter, accused the Russians of attempting to destabilize countries around the world by sowing disinformation about coronavirus vaccines on social media. He described this as a form of political warfare intended to promote anti-vaccination groups' cause. In his view, states like Russia and China 'see the strategic context as a continuous struggle in which non-military and military instruments are used unconstrained by any distinction between peace and war'. Their goal is to win without going to war.[1] More recently, the United States and European representatives on the UN Security Council accused Belarus President Alexander Lukashenko of using migrants drawn from the Middle East to create a crisis within neighbouring Poland as an act of revenge for the imposition of sanctions. Although not directly involved, Russia has demonstrated its support for Belarus by sending nuclear-capable Tu 22M3 long-range bombers on training missions over the area.

Recognizing the link between political warfare and how it reinforces war in the real world is essential. The classic illustration of this synergy was the Russian military intervention in Ukraine and Crimea in 2014. The use of non-attributable forces and support given to local proxies maintained by a political and information campaign confused and delayed an effective Western response, and ensured that basic strategic concepts such as deterrence were challenged by the range and sequence of kinetic and non-kinetic actions initiated by the Russians. This low-level form of warfare has been described as hybrid war, ambiguous war, sub-threshold war and grey zone war. The last of these terms seems to be the one most commonly used to describe

---

[1] General Sir Nick Carter launches the Integrated Operating Concept at the Policy Exchange, Westminster, London, UK on 30 September 2020: https://www.gov.uk/gov ernment/speeches

the kind of military competition that has emerged between major nuclear powers. This is broadly defined as:

> an operational space between peace and war, involving coercive action to change the status quo below a threshold that, in most cases, would prompt a conventional military response, often by blurring the line between military and non-military actions and the attribution for events. (Robinson et al, 2018, 8)

The key features of this form of war are as follows. First, grey zone activities remain below the threshold that would justify a military response. The aim is to avoid triggering military activity by either the United States or the target. As a result, the desired result is to produce a series of actions that are non-attributable. It is also possible the belligerent will initiate a cycle of violence that spikes and then falls off before a response can be mobilized, followed by a period of peace but then followed by another spike in violence. The second characteristic of gray zone war is that it tends to unfold gradually over time. These tactics might be initiated over many years and this denies the defender the ability to respond decisively. A third component of the grey zone is the extensive use made of legal and political justifications, based on historical claims and supported by documentation. Fourth, grey zone campaigns often do not threaten key vital national interests. This weakens the case of extended deterrence. 'An important quality of grey zone campaigns, therefore, is that they reflect a long series of limited fait accomplis' (Tarar, 2016). Fifth, grey zone campaigns target specific weaknesses in the target state. These can include political polarization, social cleavages, economic stagnation, the grievances these cause, and a lack of military and paramilitary capabilities. Finally, the belligerent uses the risk of escalation as a source of coercive leverage.

The doctrinal response to grey zone war

The big question the American and British militaries have been trying to answer since 2014 is how precisely to address such a nuanced and subversive form of war. Technological change has caused a profound shift in the military's perception of the battlespace both in temporal and geopolitical terms. The increasing power of computers and the information architectures used to transmit power have created new vulnerabilities which focus on the cohesion and morale of the state and society as well as the forces representing that state in a theatre of operations. Ensuring the proper utilization of force requires that its use should not be constrained by the subtle actions carried out by an opponent in terms of political warfare against the electorate and the government. This has required the military to expand their understanding of war beyond the traditional domain of the military. This expansion can be seen in two important developments. The first is the militarization/securitization of the cyber domain, which is intended to stop the flow of propaganda from an adversary, especially one as organized as in Russia and China where literally armies of trolls pump out disinformation which is intended to confuse. The second is the belief that war is now a persistent competition and can range from covert to overt actions, including large scale military operations. As such, the morale and resilience of the nation need to be buoyed up by the government's own political warfare campaign. This extends across the border of the homeland and into areas and countries overseas where competition for control is contested between democratic and authoritarian forms of government seeking to expand their influence (UK Ministry of Defence, 2020, 4–7). Greater resilience is not militarization but focuses on ensuring electorates are presented with the truth so they can see why an action has been deemed necessary by the government. It also means that, if other forms of coercion are applied which impacts on the day-to-day lives of people, they can understand why they need to endure. A good case in point is the way in which the Russians have used Europe's dependence on Russian oil and gas to break the will of NATO governments to end their support for Ukraine's war against Russia. Such an action assigns an increasingly important role for the state in terms of creating national resilience through a range of institutions.

In summary, the weaponization of information is a key aspect of the changing character of war and will have a profound impact on state and society in the Western world. It is important to acknowledge that, in contrast to previous military revolutions, the wider implications of this change in the character of war have been recognized by the military. Most important is the realization that non-military measures have become increasingly important in the conduct of war, and this requires a strategic construct that recognizes the importance of this. In the past, the military has been heavily criticized for failing to understand the broader political context within which force is used, but on this occasion, the militaries in both the United States and UK seem to have understood the need to address the challenge posed by war in the grey zone, which means looking beyond the battlefield and identifying how the enemy might exploit other vulnerabilities.

The return of strategies of deterrence

Despite the many challenges posed the Western approach to state-based adversaries remains one of 'deterrence vice coercion' (DCDC, 2021, 3). This extends beyond defence to something more dynamic and all-embracing in which all the levers of national power – diplomatic, information, economic, financial intelligence police and military – are integrated into a combined strategy. Within this competition, the role of military power is to deter and if deterrence fails, fight and win a war against an enemy. As the US military observe, 'the surest way to prevent a war is to be prepared to win one' (US Army TRADOC, 2018, 5). An important enabler in achieving this goal has been the exploitation of US technological innovation as a way of reinforcing deterrence and creating decisive advantages within the context of a future war.

This rather orthodox view of how best to secure the state from current and future security threats preserves the war–state relationship that existed during the Cold War. Of great importance here is the reliance placed on the exploitation of technology in sustaining deterrence and strategic stability. This requirement feeds back into grand strategy as technology has once again assumed a central position in creating an effective deterrent on the battlefield (US Government, 2018, 3).

In the case of military power, the aim is to 'create armed forces that are both prepared for warfighting and more persistently engaged worldwide through forward deployment, training, capacity building and education. They will have full spectrum capabilities – embracing the newer domains of cyberspace and space and developing high-tech capabilities in other domains' (HM Government, 2021, 22).

When operating below the threshold of open conflict, the role of military power is to demonstrate the state can defend itself if attacked, which means having the physical capability to protect the state from direct attack from land, sea, air and space. It also means being able to prevent a surprise coup de main. Moreover, if conventional war happens the state must possess the wherewithal to penetrate the enemy's anti–access and area denial systems (A2AD), which refers to missile defence. This means destroying or neutralizing the array of long-range precision weapons and the supportive network of sensors employed to create an elaborate defence in depth. It must also possess the resilience to protect its command and control architectures in physical and virtual space to be capable of strategic and operational manoeuvre. Finally, the United States must be able to sustain and generate forces in the theatre of operations despite the enemy's long-range fires.

In competing with an adversary in a contested geopolitical space such as the South China Sea or Russia's near abroad, Western forces rely on the presence of their forces within neighbouring countries in this space. These forces train openly to demonstrate their military potency, which it is believed will deter enemy action. The forward presence of forces is also used to contest the enemy's unconventional warfare capabilities employed below the formal threshold of war. At this level, their role is to increase the resilience of local allied forces and ensure indigenous military units can stymie the enemy's actions. Also important is the exploitation of the information domain to challenge the enemy's political narrative and space, and cyber to gain intelligence and, where necessary, to attack the enemy's irregular forces (US Army, 2018, 28). What happens if deterrence fails and the West needs to demonstrate escalation dominance because they have the means to defeat the military? How do analysts visualize a future battle incorporating the latest and future technological capabilities?

The military vision of the future battle space

The latest thinking on forthcoming change asserts 'that the future operating environment is unlikely to feature decisive battles (there will be no Waterloo moment)' (MOD DCDC, 2021, 35). The objective is to bring the conflict to an end in the quickest way possible and return to the preservation of the status quo as it existed before the eruption of violence. Within this context:

> The identification and application of de-escalation 'off ramps' from combat will be a feature of the information battle by raising the cost to the adversary of continuing armed conflict to an unacceptable level.
>
> De-escalation will be part of the continuous strategic campaign of returning the challenge to a favourable sub-threshold status which can endure. (UK Ministry of Defence, 2021, 35)

To deter war means being able to demonstrate that you can fight in a future battlespace.

As articulated in the concept of MDI, a cursory glance at the US vision of future war provides a detailed manual on how to fight a battle with tomorrow's technologies. This forms the principal document of the US military's thinking on the subject, which they acknowledge and justify in the following terms: 'The US Army in Multi-Domain Operations is an operational level military concept designed to achieve US strategic objectives articulated in the National Defense Strategy, specifically deterring and defeating China and Russia in conflict and competition' (US Army, 2018, 24).

The key term in this statement is deterrence. The position taken by the military is that once you transcend from grey zone to open war, you need to demonstrate you have escalation dominance, which means being able to win the battle.

If the United States chooses to breach the unwritten code of conduct implicit in grey zone war and opt for open warfare, then MDI is intended to provide the conceptual means to allow it to engage and defeat the enemy through a surge of military activity. This attack's intellectual and practical focus is on trying to unravel the Gordian knot created by advanced technologies, which have resulted in the creation of an expansive and complex system of defence.

This is organized in layers, based on the ranges of weapons and exploits the fruits of the latest technological revolution in the form of machine learning, robotics, hypersonic missiles, electronic warfare and the weaponization of space by both the Chinese and Russians. The aim is to create a long-range killing zone fused with an array of sophisticated sensors that are designed to prevent US forces from coming to the aid of allies on Russia's border in Europe or those states facing the threat of military action from China in the Asia Pacific Region, specifically the contested area of the South China Sea.

Current technological change envisages future battles being waged over vast distances, dominated by levels of precision strike exceeding those employed in Gulf Wars 1 and 2. Compounding this lethality is an increasing reliance on the use of autonomous weapons and systems reliant on AI to loiter over the battlefield until a target is identified and then attacked with ruthless efficiency. Latiff (2018) explores the idea of a battlefield dominated by machines. The most dramatic aspect of this visualization is that, via the development of AI, machines will increasingly think for the commander (Latiff, 2018, loc 294). In some of these projections, the speed of decision-making initiated by a computer in terms of the see, think, act paradigm is compressed into nanoseconds, making the role of the human in the chain of command all but superfluous (Scharre, 2018). Important functions within the headquarters of the division, corps or army will be stripped down as AI takes on responsibility for the processing of vast quantities of data used to understand and anticipate the enemy's likely course of action; thus addressing the age-old problem of information overload and the breakdown of cognition caused by the frailty of the human mind, which plagued commanders in the past (van Creveld, 2020a). Quantum computing will result in dramatic improvements in computing power, which will impact the speed and efficiency of radars and other sensors. In addition, it will offer a more precise form of navigation and targeting than is currently available via GPS. Equally important, it will also result in more accurate satellite imagery, and it could revolutionize cryptography, making it possible to break any code used by an enemy. It is believed that it will also allow the identification and location of enemy weapons systems that can, currently, be hidden from view, such as submarines.

Retired US army general, Robert Scales (2019) asserts that emerging technologies will not only result in unprecedented automation on the future battlefield but, most importantly, also the lethality of the weapons employed will create something similar to the bloody stalemate confronted on the Western Front during the First World War. The principal difference will be one of scale as this kill zone is raised from the tactical space of the forward edge of battle to a vast area of literally thousands of miles of territory. Such an increase in the range and accuracy of firepower tends to reinforce the power of the defence. On paper, at least, the latest iterations of A2AD pose severe problems for the attacker in terms of crossing this new zone of death laced with air and ground–launched guided missiles and artillery linked by computer-controlled radars. At the same time, Western capabilities will be eroded by enemy employment of cyber and electronic warfare, which is intended to jam all information passing between Western allies and their forces. For these reasons, MDI seeks to exploit technological innovation, expand the domains of war from three – land, maritime and air – and seek even a temporary advantage via the exploitation of space, cyber, electronic warfare and information. For Scales (2019), as in the case of the First World War, tactical dilemmas created by long-range precision strikes on a scale not experienced before are shaping strategy. The critical challenge is to avoid a new version of trench deadlock that is elevated to a larger geographic space, in which traditional models of conventional war, captured in the blitzkrieg, are obsolete and, if employed, will be suicidal. As in the case of the trench deadlock, the challenge is to devise the means to defeat the enemy's A2AD defence, which has now created a no man's land of over 100 miles across rather than a mile or so as in the First World War. MDI primarily focuses on how to achieve this task. The need to secure freedom of movement between the home base and the theatre of operation is fundamental to success on the battlefield. This means keeping sea lines of communication and air bridges open to concentrate the necessary reinforcements. Stopping the enemy from intercepting these reinforcements will rely mainly on the employment of space, aerial and electronic surveillance to locate where their central systems of command and control and launch sites are in this elaborate defensive system. These systems will then be subjected to electronic and physical attacks by ground, air and sea-based forces. The extent to which Western forces can secure their satellite platforms from attack in space is of fundamental importance (Scales, 2019).The potential expansion of the war into space is becoming more likely simply because Western reliance on satellites has increased dramatically. So, for example, one reason the US military was able to surprise the Iraqis in the ground war in 1991 was that they believed that to navigate the Americans would have to rely on the terrain of the Wadi al Batin to move their forces. They failed to realize that the US forces used GPS and so were able to move across the open desert without reference to significant terrain features. This allowed the huge left hook carried out by VII Corps across the featureless desert to bring about the encirclement of all Iraqi forces in Kuwait. Similar problems occurred at the tactical level as well. For example, in the Battle of 73 Easting, the 2nd Armoured Cavalry Regiment defeated a brigade of the Iraqi Republican Guard and elements of the 10th Armoured Division. The Iraqi commander, who had been trained and educated at the US Army command and staff college and so was familiar with US tactical doctrine, deployed his forces to bring maximum firepower onto the road, which he believed the Americans would have to follow if they were to navigate as they moved through the Iraqi desert. However, using GPS, the armoured brigade attacked from an unexpected direction and caught the enemy off guard. In the resulting battle, the Iraqi force was demolished. Over 50 Iraqi tanks and 25 armoured personnel carriers were destroyed in a 23-minute engagement. The Americans suffered no casualties (McMaster, 2016).

In the current geopolitical setting, both China and Russia will seek to challenge and negate the ability of Western forces to use its satellites to concentrate the manoeuvre of air, ground and maritime assets in a theatre of operations and allow the coordination of movement and precision-guided attacks against the enemy. Blinding US military forces by knocking out satellites is a critical objective of all state-based militaries in times of war. Control of space is as vital to modern military operations as control of the air has been since the battle of the Somme in 1916. Denial of space to an enemy can be achieved through various means, for example, electronic jamming or, in extremis, the use of electromagnetic pulse. Protecting space-based assets is problematic, but these systems are clearly increasingly vulnerable to attack.

There is concern that the United States is rapidly losing its technological advantage in space. China is pushing a state-led space programme, and the Russians have also relaunched their space programme to contest control of this domain. Of particular concern is the development of Russia's anti-satellite capability (Bitzinger, 2016; Erwin, 2018). Denial of space to the enemy can be achieved through various means, for example, electronic jamming, lasers or, in extremis, nuclear weapons. Satellites in low Earth orbit are incredibly vulnerable to this form of attack as they are located at altitudes that can be reached by ballistic missiles when launched.

The deployment of a space-based anti-missile defence system would be prohibitively expensive, and airborne capabilities such as lasers to intercept missiles are still technologically immature. At the same time, protecting one's own space-based assets is vital. It is an established maxim that in war, it is vital to establish control of the air before ground and maritime forces can move. It is now a requirement that we control near space before commencing with an attack on the planet's surface and that forces should be on the defensive until control of the space domain is achieved. Consequently, the defence of space-based assets is vital if war is to be waged effectively. However, traditional means of defence, for example, the deployment of a space-based anti-missile defence system, would be prohibitively expensive, and airborne capabilities such as lasers to intercept missiles are still technologically immature. One possible novel solution aired in US Army circles is the idea of deploying a particular type of satellite. Called Power Star, this project is based on the use of vast balloons, measuring as much as five kilometres in diameter, built of a unique fabric, with the solar panels, circuitry and computing power incorporated into the material using 3D printing. This could be launched relatively quickly from a rocket fired into space. It is claimed that such a system will be virtually indestructible once in space. Lasers or missiles would not be able to inflict severe damage due to the localized nature of the power configuration. It is also claimed that such a system could be equipped with lasers, and one such satellite in geostationary orbit could transfer as much as 50 megawatts of energy to as many as 50 military bases in the Middle East and western Pacific (Brown, 2018, 125). A recent example of space warfare was demonstrated in November 2019 when Russia launched the satellite Cosmos 2542. Once in space, this satellite then released a smaller satellite, 2543. Several months later, this system then fired a high-speed projectile. It was inferred that this was a weapon being tested to see if it could destroy enemy satellites (*The Economist*, 2020).

There are more accessible ways of achieving this goal; for example, jamming the signal sent from the satellite provides a cheaper and technically easier option. Again, this strongly implies that the war in the electromagnetic spectrum will be of fundamental importance if land, air and maritime forces operate effectively. This is only one piece of a layered intelligence network which includes fifth-/sixth-generation fighters, drones, electronic monitoring, the use of special forces and human intelligence, all of which are combined to triangulate the location of air defences, short-range ballistic missiles and long-range multiple rocket launch systems. The enemy will also attempt to degrade these capabilities by using jammers and dazzlers, and camouflage.

It is also essential to recognize that the enemy will possess these capabilities, which significantly complicates the challenge facing the attacker, which is deemed likely to be the West. On the future battlefield, technology favours the defence. What will this mean in terms of how forces are used on the future battlefield? According to Scales (2019), increased accuracy and lethality on the battlefield will see the elevation of what used to be the empty battlefield across a wider space with its size dictated by the range of long-range precision weaponry. Within this space, survival is achieved by dispersal small mobile units (approximately the size of a US Marine Corps section, 13 men) over a wide area that, it is hoped, will be challenging to hit. These forces will be physically detached from the rear areas of support. In contrast to past battles characterized by linear formations in which vulnerable flanks are protected by adjacent units linked physically or via firepower, these smaller teams will also be isolated physically. Operating in isolation for an extended period produces a range of problems. The first and most immediate is how much equipment will these small units have to carry. Currently, soldiers carry nearly 100 pounds of kit, but they do not fight effectively if they are forced to carry more than a third of their body weight. The application of technology can help here, for example, the use of exoskeletons that improve the strength and speed of soldiers. Robotic vehicles can also carry supplies.

In addition, drones can provide resupply. A critical problem created by the empty battlefield is how this saps soldiers' morale and cohesion. Small teams may also become vulnerable to this condition on the future battlefield. In this context, some form of ruggedized social media facility might help to sustain these combat teams scattered across the battlefield.

The function of these small units on the battlefield will also change, instead of engaging and winning the fight at the forward edge of the battlefield. 'In the future, small units will become virtual outposts, in effect the eyes and ears and probing fingers of a larger supporting operational force out of reach of the enemy's long-range fires' (Scales, 2019). These small forces will use sensors utilized by special forces today. Most importantly, they will have access to combat capabilities three or four levels above their normal level of command, which translates as battalion and brigade level assets employed by sections. The use of AI will also allow a complex decision-making cycle on the employment of long-range fires to be removed. This will apparently be replaced with a small unit app that will allow the strike to be carried out in seconds rather than hours. Other apps will also be available for intelligence and resupply.

It is inferred from this description that larger ground formations will remain out of range of the enemy missile, rocket and artillery fires until they have been disrupted or destroyed, allowing the deployment of this combat power onto the battlefield where it can be used to facilitate the continued destruction of the enemy A2AD system of defence.

Within this scenario, there are conflicting demands to miniaturize capabilities, but, at the same time, there is also an increasing demand for greater power from these systems. These objectives are being reconciled to an extent through the use of AI. Using this technology, it is anticipated that a rifle will achieve unprecedented accuracy, equating to one shot, one kill (Scales, 2019).

An obvious solution to the challenge posed by the increasing lethality of the battlespace is to devise ways to overwhelm the enemy defence. One way of doing this is through swarms of autonomous machines in the form of drones. The utility of armed drones was clearly demonstrated in the recent war between Armenia and Azerbaijan in 2020 and in the Russia–Ukraine War, which started in 2022. It is believed that the use of Turkish drone technology by the Azerbaijanis was a key reason why they prevailed in this short war. Drones accounted for the destruction of hundreds of armoured vehicles and even air defence systems. In some cases, the drones provided target location information for artillery or rocket systems to attack the enemy. On other occasions, the drones launched missiles or rockets at the targets. The relatively low cost of the technology employed was particularly important in this conflict. The Turkish TB2 drone costs approximately US$1–US$2million per unit compared to the US$20million paid for UK drones. The cost of the drone is so low because it uses a commercially available Garmin navigation system. The possibilities created by using cheaper drones have led to UK efforts to develop their own version and allow the creation of mass.

The next evolution in drones focuses on their use in swarms. In the civilian world, elaborate displays using drone swarms have been employed to celebrate New Year in London in 2021 and, more recently, the closing ceremony of the Tokyo Olympics. In these settings, coordination and control is exercised by a computer on the ground, which orchestrates the light show. Currently, most drones rely on a human operator to control them. However, such a system of command and control is unlikely to work in times of war as the operator needs to be in close range with the drone swarm, and the enemy will seek to jam the signal from the operator to the drones. In addition, reliance on commercial Wi-Fi technologies to allow communication between drones means swarms are generally limited to a hundred units. Using existing defence communications systems might allow this to be increased to a thousand units, but this fusion is still being tested.

This is why it is important that each unit has its own limited AI capability that allows each drone to operate autonomously. This capability already exists. The Israeli Harpy unmanned combat air vehicle is an autonomous drone that loiters while searching for targets, which it then attacks without reference to a human operator. The Chinese ASN 301 does precisely the same thing (Scharre, 2018).

The capability exists, but the question is how to expand this so it is incorporated into hundreds, if not thousands, of drones. Suppose successful autonomous drone swarms will be able to coordinate their activity with the others in the group. Such a capability can act as a decoy and suppress air and ground-based defences by electronically jamming radar or employing some form of munition to destroy them. This is in addition to the established role of drones in providing surveillance. This technology is still in development, but its potential has been demonstrated in experiments carried out in the United States and China. The UK is also investing in the development of this capability, and in 2019, the Ministry of Defence invested £31 million in mini drones including a drone swarm. It also awarded a contract worth £2.5 million to a company to develop drone swarming technology. Massed drone attacks are not new, and a smaller scale version of this type of attack was used to attack the Saudi Abqaiq oil processing facility and the Russian airbase at Kheimim. However, an autonomous drone swarm using hundreds or even thousands rather than 10 or 20 drones has not yet been unleashed in the real world. Such a capability might play a vital role in covering the advance of ground forces on a future battlefield and allowing them to close in with the enemy. Drone swarms can operate in a dispersed pattern over the battlespace in search of targets and then rapidly concentrate on one or multiple targets simultaneously. It is also feasible to use drone swarms to shield the advance of manned fighters and naval combat vessels. The critical challenge to the effective operation of drone swarms will be the enemy's ability to camouflage and hide. Within this context, the human-machine interface offers a solution in that human vision is still superior to robot systems of vision that use algorithms to detect and classify objects on the battlefield.

The possible use of drone swarms highlights its potential to weaken an enemy force by targeting critical vulnerabilities in its infrastructure, for example, attacking its logistics and supply capability behind the front line and the enemy's command and control in the form of headquarters. Ideally, these attacks would engage multiple targets simultaneously to ensure maximum physical and psychological shock.

Swarms can also be used to reinforce the defence. For example, unpiloted underwater vehicles could be widely deployed to protect vital ports from attacks from the sea. The presence of such a diffuse threat could present a huge challenge to maritime forces seeking to project military power onto the land domain (O'Hanlon, 2018).

Machines will also be the principal agents of destruction on the battlefield. Their endurance, speed range, ability to act autonomously and precise lethality make them an ideal replacement for the human on an increasingly lethal battlefield. As a result, the speed and tempo of battle are expected to increase not simply because of the advent of AI but because of the introduction of such things as hypersonic missiles. This is not an entirely new technology; for example, intercontinental ballistic missiles travel at hypersonic speeds, but they follow a parabolic trajectory that takes them out of the atmosphere before descending, which makes it easier to detect these weapons. The target state will also have longer to respond.

In contrast, developments in engine technology and the development of materials capable of withstanding extreme temperatures allow hypersonic systems to achieve speeds of Mach 5 or more in the atmosphere, follow the nap of the Earth and can alter their flight path. As a result, these weapons are harder to detect, and there is less warning time. Most important, existing ballistic missile defences cannot engage hypersonic missiles, which has important implications for the offence defence balance and overall regional and global strategic stability. Currently, only a few states, principally China, Russia and the United States, either have these weapons or are developing them.

An important innovation in the future battlefield is the likely introduction of directed energy weapons. The impetus for developing and deploying this capability is mainly in response to the threat posed by automated systems on the battlefield and the likelihood of autonomous drone swarms becoming a common feature of modern war in the future. Hypersonic missiles are another weapon that lasers might be able to deal with. The principal challenges confronting the deployment of this capability on the battlefield are access to a sufficiently large power source and storage of energy to be used.

The spark of AI animates the foundation of the technologies that will define the future battlefield. There is a tendency to exaggerate the impact of AI, and indeed it is unlikely we will see the emergence of anything on the scale of general AI, a format that comes closest to replicating the human mind's capacity. Instead, AI will be confined to what has been termed narrow AI, capable of operating autonomously to complete a limited range of missions like winning a game. This includes a capacity to memorize and learn, but narrow AI operates within clearly defined rules that are based on probabilities to select a course of action that offers the greatest chance of success. Nevertheless, this unconscious intelligence will come to be a vital force multiplier in both war and battles of the future. It is believed that the US military concept of MDI, which seeks to exploit the fleeting openings created in enemy A2AD defences, will happen at such speed that human decision-making will need to be augmented by AI to ensure the observation, orientation, decision and action cycle, (known as Boyd's OODA) loop will be fast enough to concentrate the required force and exploit the gap created (US Army, 2018).

Jamming of communications will compound the speed problem and emphasize the importance of autonomy for humans and machines. Currently, because of the limitations of AI, it requires human judgement to ensure that a broader context shapes the decisions that human experience and understanding bring to bear. This means that we are not going to see the emergence of AI-dominated chains of command anytime soon, even though, on paper, delegation of control to AI would result in an OODA loop that was completed in nanoseconds rather than minutes and hours. The potential limitations of the current state of AI in the military realm are illustrated through the US Air Force's Project Maven. This is an AI algorithm developed with the help of Google, which pulled out in 2020 and then Microsoft and Amazon stepped in to create a system capable of sifting through masses of data to identify targets across a large geographical area. Machine learning was tempered by the application of human intervention to correct the many mistakes made by the algorithm. The sole aim of the algorithm is to speed up the target acquisition cycle, an essential capability in a potentially fast-moving war. AI will be exploited within military headquarters to accelerate the decision-making process and ensure commanders are not overwhelmed by the sheer volume of data being gathered by various sensors in space, in the air, on the surface and, in the case of the maritime environment, below the surface. The current military doctrine also assumes AI-enabled weapons will play a vital role in the prosecution of battle. Most significant will be the deployment and use of lethally autonomous systems (LAWs), essentially AI-armed robots, with the discretion to use force based on the parameters of its programming. While this aspect of future war poses cultural and ethical concerns, this capability already exists, albeit in a defensive form. The Phalanx anti-missile system employed on naval vessels or the US Army's counter rocket, artillery and mortar systems have a certain degree of autonomy. In the future, many more weapons platforms will likely have this automated defensive system built into them as the speed of the attack increases beyond the ability of humans to respond. Indeed, this is one of the salient characteristics of modern A2AD defensive systems (Cox, 2021). Ethical and practical considerations will ensure that the future battlespace will be a fusion of human and machine intelligence. The logic of this position is not merely a whimsical desire to keep the human at the centre of war but a recognition that the strengths and weaknesses of both can be blended to achieve the most effective combination. So, in the air, the survivability of piloted air systems is being addressed, creating aircraft capable of being flown by a pilot in the cockpit or autonomously. This is one of the design parameters behind the UK's planned sixth-generation fighter, Tempest. In the United States, the Department of Defence has experimented with pairing a piloted F-35 with an unpiloted F-16 which is able to fly on its own. In the US Army, efforts are also being made to employ the piloted AH 64 attack helicopter with drones which will be used in environments that are deemed to be too threatening to a piloted system. In this setting their primary function will be to provide reconnaissance and target acquisition.

Conclusion

The current military vision of future war is based entirely within an orthodox geopolitical setting dominated by great power competition and focuses largely on the traditional roles of military power which are to deter and, if necessary, fight large scale battles against other states. Within this competitive space, technology has come to assume an unprecedented importance, which reflects trends within wider society. As the UK government's latest defence review observed 'technological developments and digitization will reshape our societies, economies and change relationships – both between states, and between the citizen, the private sector and the state. S&T will bring enormous benefits but will also be an arena of intensifying systemic competition' (HM Government, 2021, 24). Consequently S&T are now a metric of power. 'S&T will be of central importance to the strategic context: critical to the functioning of economies and societies, reshaping political systems and a source of both cooperation and competition between states' (HM Government, 2021, 30).

But the implications of why and how technology will shape politics is not really addressed beyond the acknowledgement of it empowering non-state actors. Seen in this light, this view confirms the argument that militaries are innately conservative and their own internal logic rather than the external environment they face shapes their theoretical and doctrinal responses to the threat. It is, however, possible to challenge this view. While the emphasis in MDI is largely on how to prosecute Clausewitzian battle in this new high-technology space, it is important that we also acknowledge that it does address the political context of conflict and war. The nature of grey zone war, with its emphasis on achieving political goals short of open war via non-military measures, allows hostile governments to sow internal division within Western states and between them. Through this action, they can cause indecision and paralysis within the Western Alliance which gives a belligerent the political and military space to initiate actions and ensure their goals are achieved without facing the risk of a general war. The first element of MDI then is to compete with the enemy within this grey zone and contest for control over this covert political and military space.

As important is the expansion of war. In essence, China and Russia have blurred the distinction between peace and war in doing so they have taken it beyond the traditional domains of land, sea and maritime forces to now include cyber and space. The first is problematic because it entails little actual physical violence and the second has largely been governed by a general understanding that the militarization of space and competition for control of this domain should be avoided in the interests of humanity, but it seems unlikely that the neutrality of this sanctuary will be preserved. The implications of this latest technological revolution ensure that the war–state relationship remains very much alive and well. Indeed, the British translation of MDI seeks to assign a central role in science and technological innovation for both defence and wider economic reasons, which sustains the war–state relationship established during the Cold War. Equally important, MDI seeks to expand the role played by all state institutions in what it describes as persistent competition both at home and overseas so that the multi-faceted threat created by technological opportunism can be addressed.

Testing Western Military Thinking about the Future of War: Russia's War in Ukraine

Introduction

One of the principal arguments made in this book is that the shift from a model of war based on the use of mass to a more capital-intensive form of warfare focused on precision changed but did not end the war–state relationship. Most importantly, the state played a vital role in facilitating this transition and remained crucial in creating technologies that changed the face of war and facilitated a broader revolution in economics, politics and society. These technologies have evolved and grown in the commercial sector and are now feeding back into the domain of war. However, brand new technologies are also emerging, and the state is facilitating the development of capabilities such as AI, quantum computing and encryption. In the previous chapter, we explored how Western militaries are engaging with the challenges posed by this latest technological revolution through MDI, this sets out the Western view of the future of war. This construct provides a conceptualization of war that widens our understanding of the subject and blurs the distinction between war and peace. An essential consequence of this reconceptualization is that it also assigns the state a more prominent role in its efforts to contest war in the grey zone. This means the state continues to fulfil its principal function as the primary provider of security.

The aim of this and the following chapter is to question the ideas underlying the current Western theorization of war. This goal will be achieved in two ways. This chapter takes advantage of the opportunity to explore the Russian invasion of Ukraine in 2022 to determine how well expectations set out in US and British versions of MDI coincide with the unfolding reality of this war. As important is the need to explore the implications of what has happened in this war in terms of the role played by technology and to ask if its application challenges or undermines Western military conceptions of future war. Within this context we also need to ponder what this war reveals about the nature of the war–state relationship that might unfold in the future. In addressing these questions the chapter focuses on the first six months of the war and so does not cover the Ukrainian offensive in autumn 2022 or Russia's use of drones and missiles to attack Ukraine's electricity grid in the hope of breaking the morale of Ukraine's citizens, which began in October of the same year. The reason for this cut-off is simple – at the time of writing, existing detailed studies cover the war from its start until July/August 2022.

Chapter 7 will then explore longer-term trends and investigate the concept of MDI against a less orthodox vision of the future strategic and operating environment.

Before proceeding further, it is essential to acknowledge the hazards of extrapolating from a single case study in this way. This is a habit we fall into too often in the world of defence and security, and it stems mainly from the infrequency of major wars in which belligerents have access to the latest technologies. Inevitably we suffer from a tendency to simplify cause and effect and overemphasize the importance played by equipment and capability. This was certainly the case in the immediate aftermath of the Yom Kippur War in 1973, which represented the first large-scale use of precision-guided missiles on the battlefield. This flawed thinking was repeated in the 1991 Gulf War when ISTAR (intelligence, surveillance, target acquisition and reconnaissance) as a concept was demonstrated in a full-scale multi-domain battle space (Toffler and Toffler, 1993; Bolia, 2004). Both conflicts were heralded as the start of a military revolution. However, these claims were subsequently challenged as new information emerged, which provided a very different complexion on the role played by new weapons and other technologies. As such, the conclusions aired here are tentative.

It is also essential to reflect on what precisely is new about the Russia–Ukraine war. According to Galeotti (2022), current military thinking on future war is essentially reinventing the wheel. He explains that today, 'war is increasingly outsourced and sublimated, fought as often through culture and credit, faith and famine, as a direct force of arms' (Galeotti, 2022, 19). So, in his view, war has not changed profoundly, and historical comparisons can be made between past and present warfare. One such period he selects is warfare in Renaissance Italy, which carries all the hallmarks we associate with the latest military revolution that focuses on grey zone warfare (Galeotti, 2022, 24).

While the historical pattern of war has displayed all the elements in the way Galeotti suggests, I believe the emphasis on each has varied in importance across time and space. As a result, it is possible to discern distinct forms of warfare shaped by the dominant technical, political and economic order that existed at a particular moment. As has been said, in the case of the era of modern war, the salience of non-kinetic forms of war continued to play an important role. However, their principal function was as a supporting enabler that facilitated kinetic action on the battlefield. In this space dominated by large-scale violence, the political outcome of the war was largely determined by a succession of battles, but activities like propaganda played a part. For example, reference is often made to the exploitation of political warfare by the Nazis during their invasion of France in 1940. This, it is claimed, tapped into and exploited the demoralization of French society in the interwar period and explains why France collapsed militarily as quickly as it did, surrendering after six weeks of fighting compared to its valiant efforts to fight on for four years during the First World War (Singer, 2018). However, let us remember that getting to this point required the Germans to deploy a multimillion-man army to force the decision and that the resulting combat was bloody, if short. It is also important to note that military historians have argued that propaganda had nothing to do with the collapse of France in 1940 and that German victory was a consequence of superior strategy and military technique (Doughty, 2014). Moreover, the British demonstrated the limits of non-kinetic war following the French defeat in 1940 as they attempted to use covert and non-kinetic activity to sow discord and rebellion across Nazi-occupied Europe. In this, they were largely unsuccessful. In the end, it took the combined military power of the Americans, the British and the Soviet Union through a series of bloody military campaigns to defeat the Nazis in 1945.

The advent of nuclear weapons changed the balance between kinetic and non-kinetic war, and the risk of escalation and nuclear holocaust led to a greater reliance on deterrence. Within this new strategic setting, non-kinetic forms of warfare increased in importance with other means to achieve political objectives, and this became a salient feature of war (Rid, 2020). In essence, it is more accurate to say that relying on non-kinetic forms of warfare is a central characteristic of what I have called postmodern war. Viewed in this way, there is a distinct continuity in the patterns of warfare from 1945 to the present. Is this, then, the future of war? A partial answer to this question is possibly provided through an examination of Russia's invasion of Ukraine. What then, can we glean from the largest war fought in Europe since the end of the Second World War and what, if any, significance does it provide in helping us to understand the future of war and the state? However, before proceeding, we also need to acknowledge the unique political context of this conflict.

The geopolitical context of the war in Ukraine

This war is essential in geopolitical terms to the principal actors involved. Ukraine is in a competitive space between two regional security systems, the first focuses on the European Union, and the second is based on Russia's conception of a security commonwealth, which is essentially the territorial space of the former Soviet Union (Buzan, 2003). As such, this war's outcome will significantly impact the future configuration of security in Europe as a whole, which means neither party can stand by and ignore the intervention of the other in this contested space. However, Ukraine is not a member of NATO and, as such, an invasion of its territory was never going to trigger a large-scale military response from the West, which means that the application of any form of deterrence was likely to be challenging. This does not mean that deterrence was not possible, and it has been asserted that had the United States left its military training teams in Ukraine, this would have sent a clear signal to the Russians not to attack, or if they did then the presence of American forces might have constrained Russian military action. However, the absence of any American presence on the ground allowed Russia to take more risks. Having said that, nuclear weapons have played an essential role in sustaining intra-war deterrence. Fear of escalation and the possibility of nuclear war have constrained the West's support to the Ukrainians. For example, while providing generous military and economic assistance, the Americans sought guarantees that if they provided long-range missiles to the Ukrainians, these would not be fired at targets within Russia, fearing this might trigger World War 3. At the same time, the Russians have also used the threat of escalation to stop the flow of material and other support to the Ukrainian government. Perhaps the most notorious warning was made by a close associate of the Kremlin who, on Russian television, informed 'the meddling British' that the Russian new nuclear intercontinental ballistic missile, Satan II, could remove the UK in a single blow. This confirms at least one part of current thinking about future war and the existence of a conceptual grey zone that might be caused by the presence of nuclear weapons and the constraints and opportunities they impose on strategy and operations.

Grey zone warfare also stipulates a heavy use of non-kinetic measures in the form of other levers of power to shape the broader political context of the war, and there is ample evidence of such activity by all the belligerents. The West relies heavily on non-kinetic measures, principally in the form of sanctions, to coerce and punish Russia for invading Ukraine. Most important, Russia's central bank assets have been frozen to stop it from using the US$630 billion it has in reserve currencies. Major Russian banks have also been removed from the international messaging system Swift, which will delay payments for gas and oil.

In the case of the Russians, they have exploited European dependence on their oil and gas to disrupt the unity of the Western coalition ranged against them. The European Union relies on Russia to supply about 40 per cent of its natural gas and 25 per cent of its oil to keep the EU economy running.

For this reason, oil and gas were not among the original commodities listed by sanctions; the only commitment made by the EU at the start of the war was to reduce their dependency on Russian gas by two-thirds by the end of 2022. Nevertheless, as the war dragged on, more significant pressure was exerted on the EU to address this gap in the sanction's regime. Current estimates indicate the Russians are earning over US$1 billion per day from EU payments for gas and oil, and this is being used to sustain the war in Ukraine (Prokopenko, 2022). In the case of gas, Europe's dependence on Russia and the absence of an alternative supplier have made it impossible to agree on more urgent action. The need to do something led Germans to cease supporting the Nordstream 2 pipeline from Russia directly to Germany. The EU has also agreed, in principle to ban Russian oil delivered by tankers by the end of 2022, however, by March 2023 this ban had still not been imposed. Moreover, the EU continues to receive 800,000 barrels of oil delivered by pipeline.

This war is potentially significant for another reason, which increases its salience within this study, and that is, like the West, the Russian approach to this war sought to capitalize on lessons learned from its recent conflicts and to exploit its latest technologies and operational technique to defeat Ukraine. As a result, many aspects of this war confirmed speculation about the employment of cyber operations and, connected to this, propaganda and political warfare. The Russians made a concerted effort to exploit their technological superiority in the cyber domain. In the lead-up to the war, they introduced malicious software into a communications satellite employed by Ukraine. Not only did this cause large-scale disruption to Ukraine's communications infrastructure, but it also disrupted nearly 6,000 German wind turbines and infected over 10,000 terminals used by the company that owned the satellite. Not content with the damage caused by this action, the Russians targeted the Ukrainian electrical grid in April. However, on this occasion, this attack was defeated with the help of Microsoft (ERPS, 2022). Russian propaganda was also essential to their more comprehensive strategy to disarm and disable the Ukrainian population. However, its message on this occasion failed to gain the support of Ukrainians, which suggests those tasked with this form of political warfare misunderstood or underestimated the strength of local nationalist feelings within Ukraine regarding the sanctity of its independence as a nation state (Pinkstone, 2022).

Evidence that Russia is losing the political war can be seen from the actions taken by its government. In March 2022, Putin banned Facebook and Twitter from using Russia's electronic ether, and any journalist found guilty of producing 'fake news' on the war in Ukraine now faces up to 15 years in prison. There is little doubt that this indicates Russia has lost the information war. It is not just that the message failed to resonate with Ukrainian and international audiences; as important have been the actions taken by governments to silence the platforms used by Russia to obfuscate and confuse the world regarding its actions. Twelve critical disinformation outlets used by Putin to bolster his foreign policy have been hit with sanctions in an online crackdown on false and misleading reports claimed to be instigated by Russian intelligence. In March, 2022 the UK Foreign Office announced sanctions were being imposed on the Internet Research Agency, which was responsible for generating much false information. There is evidence that this organization paid Russians US$650 per month to flood the internet with pro-Putin comments on chat forums, social networks and comment sections of Western publications. In addition, two alleged disinformation websites, New Eastern Outlook and Oriental Review, were also targeted. The US Treasury also imposed sanctions against these groups. In addition, they also targeted the Strategic Culture Foundation, a Russian think tank which commented on global current affairs, but which was accused of promoting disinformation in Western media. Similar accusations were directed against Russian media outlets such as well as SouthFront, NewsFront and InfoRos. All three were accused of having links to Russian intelligence. These groups' Twitter and Facebook accounts were also blocked to stop them from exploiting Western social media (Thomas, 2022). Western states have also implemented countermeasures to prevent or limit cyber-attacks. As a result, bots – automated systems used to create thousands of social media accounts and which have played an important role in recent influence campaigns – are being identified and blocked. In response, the Russians are relying increasingly on 'troll farms', which use people to recruit and coordinate supporters on the internet to target social media profiles of people or organizations critical of Russian policy. Using real people rather than bots decreases the chances of detection. Russia's propaganda war was further undermined because of the perceived brutality of its campaign, which lacked proportionality or discrimination in its attacks on innocent civilians, who became the victims of indiscriminate military offensives against strategically important towns and cities. More worrying was the behaviour of Russian soldiers post-conflict and accusations of the brutal treatment inflicted on the occupied population by Russian security forces.

In contrast, Ukraine has been far more successful in promoting its cause to various audiences, national, international and even Russian, which has legitimized its opposition to Russia's assault and mobilized wider support for its cause. An important part of the Ukraine and indeed the West's strategic information warfare campaign has been the role played by Bellingcat. Set up in 2014 it rose to prominence because of its work on the use of chemical weapons in the Syrian civil war. It's distinctiveness derives from its reliance on digital media and all forms of open source intelligence to conduct its investigations. In the case of the Russia Ukraine conflict it played an instrumental role in identifying what role Russia played in downing

Malaysian airlines flight MH 17 over Ukraine in 2014. More recently Bellingcat's use of open-source intelligence has legitimized claims made by the Ukrainian government regarding the treatment of Ukrainian prisoners of war by the Russian military and attacks against Ukrainian civilians. http://www.bellingcat.com/

Open-source intelligence has also proved essential in helping the Ukrainians meet the challenge of Russia's invasion. The smartphone means virtually every Ukrainian citizen checks and triangulates information on the internet and verifies and geo-locating images in real-time. It is believed that every truck and troop movement is being tracked and logged. In the past, disinformation was able to spread because there was a delay between an event and the emergence of accurate intelligence on what happened, which allowed people to manipulate the truth to achieve their own ends. The generation of open-source real-time intelligence is preventing this from happening because the information is available immediately (Cadwaller, 2022). However, perhaps the most surprising and interesting aspect of this war lay in the way in which military power was used in the physical domain. In exploring the conduct of the war the debate has focused on the extent to which technology shaped the character and the outcome of the conflict. This matter forms the principal focus of the next section of the chapter.

The kinetic dimension of the war

It is essential to recognize that what happened in Ukraine on the ground, in the air and at sea also challenge aspects of Western thinking on future war. Most surprising was the scale and ambition of the assault when hostilities commenced. Pre-war, most experts believed the invasion would be like the war in 2014, which was based on the pursuit of limited objectives using limited means and in which the balance of investment was very much in non-kinetic activity. What unfolded in 2022 sought to achieve the complete conquest of Ukraine, and the character of the war that unfolded more closely resembled the battles of the Second World War than a postmodern war. Approximately 200,000 Russian soldiers were massed along the border with Ukraine. However, in the lead-up to the war, it was believed this was part of a coercive diplomatic strategy which aimed to obtain definitive security guarantees from the West without recourse to an actual war. The Ukrainian army's full-time strength was 125,600 troops supported by 900,000 reservists. As it became clear a political compromise was not possible, the use of force looked all but inevitable, but the scale and ambition of the attack still came as a surprise. In this version of Russian blitzkrieg warfare, multiple armoured columns sought to achieve the rapid and decisive conquest of Ukraine. These were preceded by missile strikes (some 900 missiles were fired in the first month of the war) and the dispatch of airborne forces whose mission appears to have been to neutralize the Ukrainian government by fair means or foul. It is believed Putin hoped to bring about the instant collapse of resistance of Ukraine through a demonstration of Russian 'shock and awe' and present the West with a fait accompli (Johnson, 2022, 17).

The expectation in this war was that Russian heavy artillery and airpower would destroy the enemy at a distance through drones and other targeting systems before closing in with heavy armoured forces supported by mechanized infantry in armoured vehicles. Russian self-propelled guns, which are carried on an armoured chassis, can strike targets out to a range of 24–26 miles and advance quickly to support their armoured formations. Russian tanks using stabilized sights can fire while on the move and hit enemy targets over a mile away. Their speed is such that they will close with the enemy in two minutes, which gives the defence little time to respond. If enemy light infantry engages, the Russian forces can deploy their infantry to neutralize this threat. In theory, the speed, firepower and protection of Russian armoured formations should have made them all but invincible on the battlefield. Before the outbreak of war, most military pundits believed Russian victory was assured and that the war would be over in a matter of days.

Obviously, the war did not unfold that way; instead, it became a slow battle of grinding attrition. By the 61st day of the war, it was estimated the Russians had suffered 15,000 killed, more than had been killed in ten years of fighting in Afghanistan, and at least double that number wounded. The Russians also lost a lot of equipment. UK military estimates suggested the Ukrainians destroyed 530 Russian tanks, over 1,000 armoured vehicles of all types, and 60 fighters and helicopters. The Russians employed over 120 battalion-sized battle groups, but after 60 days of fighting, 25 per cent of this force was combat-ineffective (Wallace, 2022). By day 180 of the war, UK military intelligence estimated that 75 per cent of these battlegroups were broken. The Ukrainians also suffered heavily in this war. US estimates indicate that after four months of fighting, between a third to half of their equipment stocks had been destroyed in the fighting and that, in the most recent combat in the Donbas, they lost between 100–300 men per day.

The failure of Russia's military intervention to bring about a rapid victory against the Ukrainians led many in the Western media to focus on the superiority of Western military technology in the hands of the Ukrainians. This image was reinforced by the Ukrainians, as demonstrated by the numerous YouTube videos showing Ukrainian soldiers firing smart munitions and destroying Russian tanks and artillery. By March 2022, 17,000 of these precision-guided munitions had been sent to Ukraine by Western governments, and there is no doubt they played a significant role in stymieing Russia's ground war.

The level of attrition suffered by both sides in this war raised important questions about more traditional forms of military power. According to Wetzel the war in Ukraine revealed the obsolescence of conventional military power in the form of tanks and fighter aircraft. In recent wars, the establishment of air superiority through the destruction of the enemy's air defences provided the freedom to bring airpower to bear on ground-based enemy defences, and this, combined with ground assaults, all but ensured the defeat of the force unable to control the skies above the battlespace. A noticeable aspect of the recent war in Ukraine is that the Russians have not been able to achieve the vital goal of suppressing Ukraine's air defence, which has limited the role played by its air force in this war. Despite Russia's massive numerical superiority in the air, with over 3,800 aircraft, including 400 modern fighters, versus Ukraine's 98 aircraft, four months into the war, it could not achieve control over Ukrainian air space. In those four months, it lost 165 combat aircraft over Ukraine, 10 ten per cent of its frontline strength. Why, then has the Russian air force been unable to destroy the Ukrainian air force and establish air superiority over the battlefield? Part of the answer lies in the quality of the Ukrainian ground-based air defence, which is a mixture of old and new systems. However, these were deemed sufficiently lethal to keep the Russian air force grounded for the first four days of the war, which proved critical when the offensive began. The Russian air force's competence has also been questioned; its failure to suppress Ukraine's air defences has been attributed to its lack of skill to organize larger-scale air operations beyond attacks involving one or two fighters. A potentially deeper problem lies in the risk-averse behaviour of the Russians, not wanting to expose their hugely expensive and technologically sophisticated aircraft to the relatively inexpensive but highly effective threat posed by anti-aircraft missiles and artillery.

Western technologies have also played a vital but indirect role in supporting the Ukrainian war effort. One of the best examples of this is the introduction of Starlink. This is an internet service provided by a network of over 2,000 satellites connected to ground stations that provide internet coverage over a wide area; it can be accessed via a terminal, and over 10,000 terminals have been sent to Ukraine to help it maintain the internet for both civilian and military purposes in areas where existing communications networks are down. This facility needs to be seen within the context of Russia's efforts to cripple Ukraine's communications infrastructure at the start of the war. The cyber-attack on the LA-SAT network mentioned earlier is one such example which resulted in partial network failure for broadband customers.

Airpower is used sparingly, and helicopters and even drones have proved vulnerable to Ukraine's air defence. As a result, the Russians struggled to destroy enemy tanks and artillery. Had the Russians relied more heavily on drones to provide close air support, its campaign might have achieved more significant results. These systems can survive for a limited time, even in contested airspace where the enemy's air defence system is active and functioning. Indeed, drones such as the Bayraktars have engaged missile systems that should have shot them down. These relatively cheap systems provide a cost-effective alternative to piloted fighters and have demonstrated their ability to survive even on a high-intensity battlefield. On the ground, Russian armoured forces have also suffered heavily at the hands of Ukraine's army of light infantry armed with anti-tank guided weapons. These weapons employ a shaped charge which focuses the explosive energy of a warhead against armour plating resulting in its penetration of the tank. In the past, armoured forces have addressed this threat by employing explosive reactive armour, which dissipates the energy of the shaped charge and preserves the integrity of the armour on the tank. However, the development of tandem warheads means the explosive armour is first detonated, which allows the second charge to penetrate a tank's primary armour. The employment of systems designed to launch an attack against the weakest part of the tank, the top of the turret, has also allowed the Ukrainians to destroy a large number of Russian tanks (Page, 2022).

AI has also played an important but less spectacular role than previously imagined in this conflict. In an initial assessment of the impact of this technology, Goldfarb and Lindsay (2022) observe that the presence of AI was having precisely the opposite effect to the claims made in that it was slowing down the tempo of operations rather than speeding them up. AI has provided an important force multiplier that has helped Ukraine to meet the challenge of defending itself from Russia's aggression. In the cyber domain, Microsoft has used its own AI capability to help the Ukrainians deflect and frustrate Russian cyber activity. It also agreed to host all Ukrainian government data on its remote servers to protect that information.

AI has been used to facilitate the information war via social media platforms, news feeds and media reports on Ukraine's perspective of the war on and off the battlefield. AI is also being used by the commercial logistics organizations that are being used to move military and other supplies into Ukraine from all parts of the world. Western intelligence agencies are also using AI to help sift through the vast quantity of information secured via a variety of surveillance systems, which has enabled them to provide timely and effective intelligence to the Ukrainian military. One illustration of the importance of this capability are the 12 Russian generals killed so far in the war. Such losses are unprecedented in recent wars and can be attributed to several factors, including lax security, for example relying on unsecured phone lines. However, the US intelligence system has also played a part in providing real-time intelligence on the location of senior Russian commanders. In sum, the literature on AI has tended to focus on the advantages enjoyed by the attacker in war, but the Russia–Ukraine war suggests that, currently, it is the defender who has gained most from the deployment of this capability both on and off the battlefield simply because AI is making it harder for Russia to defeat Ukraine (Goldfarb and Lindsey, 2022).

The salience of technology in determining the early phases of this war can be challenged on the justifiable grounds that the failure of the Russians strategically, operationally and tactically has much to do with incompetence, providing a compelling case which explains the failure of this campaign. Rob Johnson (2022) provides a concise analysis of these flaws. He explains that the Russians failed to follow the basic principles of war and demonstrated a surprising lack of skill in the conduct of combined arms warfare. They failed to neutralize Ukraine's air defence system, again attributed to a lack of skill, and their armoured columns followed predictable routes, which allowed the Ukrainians to set up ambush positions to stop the advance of Russian armoured columns (Johnson, 2022).

It is also important to remember that, from 2015, member states of NATO provided comprehensive training to the Ukrainian armed forces. The British trained over 20,000 Ukrainian soldiers between 2015 and 2020, and the Americans trained a similar number, all of which meant the Ukrainian military was better prepared for an assault in 2022 than in 2014. The Russian failure to understand how strongly the Ukrainians would resist meant fighting was more intense and prolonged, increasing the burden on logistics. Even basic factors such as topography need to be considered. The more complex terrain around Kyiv allowed Ukrainian infantry to remain hidden from the Russians and get close enough to strike their targets. Finally, the offensive coincided with the start of the Spring thaw, which meant that Russian armoured columns were mainly confined to the roads, which frequently resulted in large queues of vehicles trying to fight their way down a single road against constant ambushes and harassing attacks from Ukrainian troops on the ground and in the air. Although the Russians were aware of the problems posed by the weather and the Spring thaw, it is believed they delayed the offensive because the government promised the Chinese government they would not launch an invasion of Ukraine until the winter Olympics was finished.

According to Rob Lee (2022), huge Russian tank losses are not the direct consequence of new technologies entering the battlefield but are due to more prosaic factors. Chief among these is that the Russians made little attempt to organize a coherent combined arms operation, instead focusing on speed and surprise in the hope they could seize their objectives with no real interference from the Ukrainians. In addition, there is evidence that the mission was rushed and the military was given minimal time to prepare. As a result, there was insufficient infantry to cover the advance of tanks, which made them vulnerable to anti-tank missiles. Failure to prepare meant that there needed to be more logistical support to maintain tanks in the field. In addition, because of the divergent axes of the advance of armoured formations, these forces often moved beyond the range of air defence, artillery and electronic warfare. The nature of the attack also meant that its logistical support was largely confined to the road; that meant it had elongated flanks and the frontage of the advance was restricted to the width of the road (Lee, 2022).

Stephen Biddle (2022) has commented on the relative competence of the forces involved in this conflict. In his opinion, the difference in combat outcomes between the Russians and the Ukrainians is not merely a consequence of technological asymmetry, the key variable in explaining battle outcomes has been the quality of the defence. Where it is well prepared then the attacker has struggled to make progress and both Russian and the Ukrainian success can be explained by the presence of this factor. As he explains:

> This should not be surprising. In fact, it encapsulates the modern history of land warfare. Since at least 1917 it has been very hard to break through properly supplied defenses that are disposed in depth, supported by operational reserves, and prepared with forward positions that are covered and concealed (and especially so without air superiority). (Biddle, 2022)

In Biddle's view, clean breakthroughs followed by exploitation and decisive conquest requires a permissive opponent 'that is a defender who lacks depth, who has failed to withhold a meaningful reserve, who has failed to ensure cover and concealment at the front, and, often, whose troops lack the motivation to fight hard in the defense of those positions' (2022).

'Often, the best single predictor of outcomes in real warfare has thus been the balance of skill and motivation on the two sides' (Biddle, 2022).

So poorly have the Russian military high command performed in this war that it was concluded that even the extensive use of AI could not make up for the cognitive failures of Russia's strategy to win this war. As Goldfarb and Lindsay (2022) observe: 'AI will provide many tactical improvements in the years to come. However, fancy tactics are no substitute for bad strategy. Wars are caused by miscalculation and confusion, and AI cannot offset natural stupidity.'

Mounting losses and lack of progress on the ground caused the Russians to pause their campaign in early April 2022 so they could regroup and reorganize. The second phase of the war began on 18 April, and it is clear the Russians learned important lessons from the fighting over the previous nine weeks. The second stage of the campaign focused on securing control over the Donbas region in the south-east rather than the entire country. The Russians also concentrated most of their military power in this area to achieve this territorially limited objective: 93 battalion battle groups were assigned to complete this task. A new commander to lead all Russian forces in Ukraine, General Aleksander Dvornikov, was appointed to simplify command and control.

This second part of the 'special operation' had the look and feel of a very different battle to that experienced during the first phase of the war. On paper, the open terrain in this area should have provided the Russians with an important advantage in conducting an offensive. The Russians also had a 15:1 superiority in artillery over their opponents and used it to good effect to suppress Ukrainian defenders and allow the advance of Russian troops towards their objectives. The lack of long-range artillery in the Ukrainian arsenal made it difficult for them to employ counterbattery fire and neutralize the Russian artillery threat. Faced by this tactical dilemma, the Ukrainians called on their Western allies to supply them with the long-range artillery and tanks needed to challenge Russian dominance in this sphere of land warfare. However, a noticeable aspect of the campaign in the summer of 2022 was how slow Russian progress was; they were winning and driving the Ukrainians back, but the pace of the advance was glacial. In September the Ukrainians launched a counteroffensive in the Donbas which caught the Russians by surprise. As a result, Ukrainian forces liberated over 1,000 km of territory captured by the Russians since the start of the war. The Russian response was to instigate a second offensive in the southern Donestk Oblast. Here fighting came to focus on the town of Bakhmut. A combination of terrain in the form of urban warfare and the onset of winter slowed the pace of war significantly and as we move into the second year of the war, fighting in and around Bakhmut more closely resembles the images of the battlefields of Ypres and the Somme from the First World War than a 21st century high-tech war. Given Russia's numerical advantage in the Donbas, especially in artillery, we need to explore why their offensive stalled.

Although the terrain in the Donbas contains good tank country, allowing for fast-moving armoured operations, it is also a landscape littered with villages all within range of anti-tank missiles, making it possible to create an interlocking defence. Fortifying these villages and using them as anti-tank positions forced the Russians to attack methodically rather than with any dash or flare and to use their artillery to destroy the defenders before pushing through with their armour and infantry. Drones continued to play a role in this second phase of the war but were used to support engagements and provide targeting and acquisition for indirect fire, but artillery seemed to be the dominant weapon system.

The Ukrainians attempted to challenge Russia's dominance in artillery in three ways. The first focused on Russian reliance on UAVs to provide real-time targeting and acquisition of for its artillery. With this capability Russian artillery could attack a target within 3–5 minutes and track targets as they tried to escape. The Ukrainian response was to deploy man portable surface to air missiles (MANPADs) to destroy Russian UAVs. The second was to interdict Russian supply lines using long-range artillery and rocket fire before ammunition arrived at the frontline and thereby reduce the number of shells that could be fired. This was Russia's critical vulnerability as the movement of supplies was fixed by reliance on the rail network, which made targeting its supply lines easier. The Russians also relied heavily on civilian cars, which proved incredibly vulnerable to Ukrainian artillery, to move supplies from the railhead to the frontline. The third option was the use of electronic warfare. Both sides have relied heavily on satellites to communicate and provide accurate guidance for navigation and carrying out precise attacks. Denial of access to this capability assisted the Ukrainians in their effort to disrupt Russian use of artillery, which has proved effective in the second phase of the battle (Watling and Reynolds, 2022, 11).

The Russians mirrored the actions of the Ukrainians in neutralizing the threat posed by drones. In the summer, they deployed more air defence systems to eastern Ukraine. This reinforcement was intended to protect Russian artillery and ensure it was available to support offensive operations in this area. It was claimed this action stopped the Ukrainians from using Turkish Bayraktar drones, which had been used to great effect in the first phase of the war. There were also concerns that US-made drones like the Gray Eagle would prove vulnerable to Russian air defence systems (Stephanko et al, 2022). The Russians also jammed the GPS signal to guide Ukrainian drones, and this too reduced their utility and effectiveness.

To some extent, the hype surrounding the use of drones has now been transferred to a new weapons system, which entered the theatre of operations in June 2022. HIMARS (High Mobility Artillery Rocket System) is a multiple-launch rocket system that fires six guided missiles with a range of 57 miles or a single Army Tactical Missile System with a range of 190 miles. It is claimed the United States has not supplied the second of these missiles because of fears it could be used to attack targets within Russia. HIMARS compensates for the loss of drones by allowing it to strike at targets deep behind the Russian frontline. It is claimed these missile strikes have destroyed Russian artillery, supply dumps, command control posts and transport infrastructure, such as road and rail bridges. The net effect of these attacks deep within Russian-controlled territory was to further slow the already ponderous pace of Russian operations over the summer and provide a counter to Russian artillery, which inflicted considerable damage on Ukrainian forces. Estimates suggest the Russians fired 20,000 shells per day against Ukrainian troops. Given this weight of fire, it is not surprising that Ukrainian infantry has struggled to hold their ground. As with other weapons deployed by the Ukrainians, the effectiveness of HIMARS is expected to decline as the Russians succeed in finding and destroying these systems, or they can jam the GPS guidance system used on the missile. Alternatively, the Ukrainians could simply run out of rockets because of limited supplies, or poor maintenance could prove to be a problem resulting in fewer operational systems.

The war in Ukraine has been compared to the First World War battlefields of the Western Front in 1915 and superficially at least there are a number of similarities between them. As in 1915, both sides have suffered huge material losses in the opening phase of the war, and they lack the means to bring the war to a decisive end on their terms. Like 1915, there is also no appetite to stop the war, and both sides refuse to compromise.

Ukraine hopes to retake all the territory lost since 2014 but lacks the force to break through the enemy's lines. Its most notable recent successes have been the launch of daring special forces raids on Russian bases deep behind the frontline. Perhaps the most dramatic of these raids was the detonation of a truck bomb on the Crimea bridge in October 2022. This attack resulted in the collapse of 900 foot span of the 11-mile road and rail bridge which joined Crimea to the Russian mainland (Glanz and Hernandez, 2022). Russia's offensive in the Donbas stalled, but they conquered a large swathe of Ukrainian territory, and, as has been said even though they were checked by a Ukrainian counter offensive, in November 2022 the Russians renewed their attack in the Donbas in an effort to secure control over the entire region. .. The Russians have also used indirect means to coerce the Ukrainian government to seek terms in this war. A good illustration of this can be seen from heir seizure of the Zaporizhzhia nuclear power station in March 2022 and their efforts to disconnect it from the Ukrainian electrical grid. This meant Ukraine lost 20 per cent of its electrical power in the winter, which the Russians knew would cause severe hardship for many Ukrainians. There were also fears this would cause a fresh wave of up to two million refugees crossing the border into Poland (Sabbagh, 2022). In fact, nearly 3 million Ukrainians left the country after 3 October 2022 and 14 February 2023 (Statista, 2023).

The Russians also hoped the support given by Western allies to Ukraine would fall away in the winter as gas and oil shortages in Europe caused widespread discontent among Europeans prompting calls to reduce financial and military aid to Ukraine. Having failed to deliver a decisive military blow and bring the war to a decisive end in 2022, the Russians are now facing the prospect of having to reorganize so they can sustain their forces in the field in the winter months, with a view to a renewed offensive against the Luhansk region of the Donbas in Spring 2023. Despite all the technology deployed in this war, it is taking on many of the character traits associated with past wars. The most important of these focuses on the challenge of replacing complex technologies lost during battle and the resort to what might be described as conventional 'broken back warfare', which envisages the continuation of the war even though both sides have lost their most prized weapons. What then then does this war reveal about the relationship between war and the state?

Implications of the Russia–Ukraine conflict for war and the state

This war challenges but also reinforces the thesis set out in this book, which emphasizes the importance of the connection between war and the state because of the presumed importance of technology in this relationship. However, Ukraine has little in the way of a modern high tech defence industrial base. Indeed, most of its industry is centred on traditional manufacturing, such as engineering and steel production and, of course, agriculture, all of which has more in common with the second rather than fourth industrial revolution. Does this then challenge or negate the war–state relationship explored so far in this book? The answer is 'no' for two reasons.

First, while we might debate the value of high-tech weaponry in this war, there is little doubt that Ukraine would have been defeated without the provision of military aid from the West, particularly from the United States. Since January 2022, the United States has provided US$13.5 billion in security assistance to Ukraine. This includes 700 single-use Switchblade 300 and 600 drones, which loiter over the battlefield until they have found a target. Another 28 nations have contributed to rebuilding Ukraine's military capability to prevent military collapse. The scale of international support given to Ukraine again highlights the importance of having access to a range of supplies in terms of military equipment, including spares and ammunition, but the need to acquire advanced weapons has been particularly important. However, it also demonstrates the political problems that arise when relying on other governments to provide the materials needed to wage war. In this case, Western governments are willing to sustain Ukraine's war effort at a level where it can continue to fight but not enough to allow it to precipitate the defeat of the Russians. Viewed cynically, this reduces the chances of escalation of a local war sucking all of Europe into World War 3, but sustaining support at a certain level to Ukraine does weaken Russia, which means it will pose less of a threat in the future. The problem is that this policy imposes a high cost on the Ukrainian people, who might become the victims of what looks increasingly likely to become a protracted war.

The solution to how to lessen Ukraine's dependency on the West for arms is quickly addressed in the long term by developing its own indigenous defence industrial base. However, given that most defence industries operate at sub-optimal levels of production because their internal defence market is not sufficiently large and the high entry costs associated with creating such a capability, this seems an unaffordable proposition for Ukraine. However, this war has revealed a possible solution to this problem that Ukraine might want to pursue in the future. One of the notable aspects of this war is the cost imbalance between weapons used by the defender and the attacker. Ukrainian forces have relied heavily on a range of technologies both on and off the battlefield to gain an advantage over the numerous Russian invaders. On the battlefield, they have relied on various precision-guided munitions to attack Russian aircraft and tanks. Most important is that some of these technologies are relatively inexpensive and considerably cheaper than the kit the invaders have chosen to use. For example, the T90M main battle tank costs about US$6 million. In contrast, the UK/Swedish light anti-tank weapon NLAW (New Generation Light Anti-Tank Weapon) costs US$37,000. The British provided over 5,000 of these to the Ukrainians. Although its effective range is limited to 600–800 metres, it has proved to be highly effective in the specific terrain of the north of the country.

The Ukrainians have also demonstrated a capacity to adapt and have blended old and new technologies to produce lethal weapons. A notable example of this kind of innovation was the use of old Soviet-era anti-tank mines. Using 3D printers, the Ukrainians could attach plastic fins and drop them onto enemy tanks. The blast can penetrate eight inches of steel plate, which means most armoured vehicles are vulnerable. Again, a striking aspect of this capability is how cheap it is – approximately US$8 for the fins and US$80 for the mine – and yet such a weapon can destroy a Russian armoured vehicle costing US$650,000. Focusing on developing industries that create dual-use technologies and manufacturing inexpensive weapons that require minimal training to use and can be produced in abundance might increase Ukraine's ability to deter future aggression. A notable feature of this war is that emerging technologies have proved most useable in defence rather than offence. This also makes military training easier than for offensive action where different combat arms must integrate and coordinate on the battlefield, which is not an easy skill to acquire, especially when your army is largely made up of a militia (Kemp, 2022).

Russia's experience of the war reinforces the perspective that their military also sees technology as a vital force multiplier. In some product areas they have few problems. Indeed, in the case of fighter aircraft, the Russian state has lost 25 fighters in the war in Ukraine, including nine destroyed in the Russian airbase at Saky, on Crimea's western coast, but they introduced 400 new fighters over the preceding decade, so it retains a highly potent capability. However, it might need help to replace the 3,000 long-range precision-guided munitions expended in the opening phase of the war. It is claimed that in all areas, efforts are being made to increase defence production, but its efforts here are being hampered by sanctions. It is also clear they did not have enough drones at the start of the war and this weapon technology was not prioritized in the lead-up to the invasion.

Consequently, an essential chokepoint in the Russian logistical chain focuses on its lack of an indigenous technology base to support the production of armaments weapons required to create an ISTAR capability. This has become an acute problem in a war where the rate of attrition dramatically exceeded expectations. Examination of Russian equipment has revealed a heavy reliance on Western military technology. According to a report from the defence think tank RUSI (Watling and Reynolds, 2022), 'almost all of Russia's modern military hardware is dependent upon complex electronics imported from the US, the UK, Germany, the Netherlands, Japan, Israel, China and further afield'. The Russians have been forced to import these components because they cannot manufacture them domestically. Based on the evidence from a crashed Russian drone, it was discovered that the Russians are not using particularly sophisticated technology but employed a camera available on the high street and other commercial technologies that can be bought off the shelf. This has become an important critical weakness for the Russians who faced with comprehensive sanctions, will now struggle to acquire these technologies (Oliver, 2022).

To address this deficiency, the Russians have re-established an industrial espionage unit called Line X. Created during the Cold War, its purpose is to steal those technologies that Russia does not have, but which are vitally important when building modern weapons. The Russians are working through third-country transhipment hubs to overcome the barriers imposed by sanctions. It is claimed that by employing these measures, the Russians have procured enough semiconductors and other technologies to supply their armed forces for the next ten years (Byrne et al, 2022).

However, as important as the need to acquire technology has been how to fill the huge gaps in manpower and more basic equipment suffered during this campaign. At the sixth month point in the war, the UK's defence intelligence estimated that 75 per cent of Russian battle groups deployed to Ukraine on 22 February 2022 were destroyed. Confronted by the reduction in combat power on this scale, there were fears that Putin would double down and mobilize the Russian nation for war. One Russian veteran of the Wagner Group estimated that it would require a force of over 800,000 troops to conquer and hold Ukraine and so, from this perspective, mobilization of Russia's military made sense. However, as has been pointed out, human resources might be readily available, but it is claimed that Putin is unwilling to call for a general mobilization for war because he fears this will spark protests within Russia. In addition, there is the practical question of how useful mobilization will be in meeting the immediate concerns of the Russian leadership. This focuses on the question of how long will it take to regenerate the capability required to renew the offensive. One Russian military analyst (Lendon, 2022) observed that a new tank division takes over 90 days to train, assuming you have the required tanks. New fighter pilots and aircraft might appear, but it will take a year to create and deploy these capabilities. There is also the cost of replacing equipment lost so far, which will run into billions of rubles. As important will be the harm inflicted on an economy already damaged by sanctions. It is expected that these measures alone will cause a 10 per cent or more economic contraction in 2022, a fall not seen since the days of economic and political transition in Russia in the early 1990s. Despite the risks involved in implementing total mobilization for war, there continued to be calls in July 2022 for Putin to abandon the pretence that this war is not a war but a special military operation and introduce conscription and place the economy on a war footing. With this additional resource, it was claimed that Russia could then successfully conquer Ukraine. Currently, Russia is raising volunteer battalions to replace its losses. These formations are about 400 strong and consist of volunteers aged 18–60. The volunteers do not require any prior military service but will receive only 30 days training before deployment to Ukraine (Hird et al, 2022). No formal declaration of war or full mobilization of manpower emerged from discussions in summer of 2022. Instead a compromise was sought in which a partial mobilization of 300,000 men was initiated in October. Expectations that Putin would declare a full mobilization for war during his presidential address to the nation in January 2023 also failed to materialize.

A possible solution to Russia's military supply problem was to draw on the vast stocks of equipment held in its stores. However, a common perception of this war is that third and fourth industrial revolution technologies have played a vital role in precipitating the defeat of older but more expensive military systems deployed by the Russians. In essence, the Ukrainians achieved a technological asymmetry, which allowed them to exploit the critical weaknesses of the Russian conventional arsenal. As a result, deploying older tanks and armoured infantry vehicles was not seen to be a viable solution but merely another way of compounding failure and was melodramatically compared to sending lambs to the slaughter. Despite these protests, evidence has emerged revealing the Russians have been forced to use old stocks of equipment to replace their losses in Ukraine. This includes the deployment of T62 tanks and old multiple-launch rocket launchers and howitzers from storage facilities in Siberia. Most worrying was the use of mines dating back to the 1950s to reinforce their frontlines in the Kherson Oblast. Apparently, several of the mines detonated while being transported to the frontline, killing an unspecified number of Russian engineers (Hird, Stepanenko and Clark, 2022).

The war in Ukraine represents a throwback to the era of modern wars, which were dominated by mass, and the pursuit of victory on the battlefield. This has been caused, in part, by the rate at which advanced technologies have been consumed in this war and this has enforced a reliance on more traditional forms of military power where quantity as much as quality of performance matters. This suggests that, even in an age of 'demassification' (as the Tofflers described war's evolution in the computer age), mass, even in a high-tech form, matters in war, and this poses important questions regarding the optimal war–state relationship going into the future (Toffler and Toffler, 1993). The war also reflects a deeper feature of war which taps into its nature as set out by Clausewitz and which highlights its social and political dimensions of war (Book 1, Chapter 1, 1976). So, for example, Echevarria (2022) has observed that this war demonstrates that large-scale conventional war is not obsolete as has sometimes been claimed. Most interesting, however, is his second observation that what we are seeing in Ukraine also reflects elements of another paradigm of war, one referred to as 'war amongst the people', a thesis set out in 2005 by Rupert Smith. Smith's principal argument was that messy, irregular wars were replacing major conventional war; a setting which made industrial age armies all but irrelevant (Smith, 2005), but this view was challenged by the likes of Colin Gray, who believed major interstate wars would persist (Gray, 2005). These two paradigms are often set in opposition to each other, but Echevarria believes that a significant aspect of the Russia–Ukraine war is that it reflects the characteristics of both paradigms. This revives a more traditional role for the Ukrainian state in this conflict, one which focuses on the role played by the state in facilitating the realization of the mobilization of the economy and society to defend their homeland. Of importance here is the belief among Ukrainians in the justness of their cause and the passion of their resistance to occupation. This has exerted a powerful impact on the shape and character of the war that has emerged in this conflict. A case in point is the failure of the Russians to seize Ukraine's second city, Kharkiv. The objective was only several miles from the border, and the population was mainly Russian speaking; not surprisingly, it was assumed the people would welcome the invasion. The Russian army also deployed their best units to secure this objective: the 1st Guards Tank Army. Expecting little opposition, they sought to seize the city by direct assault. However, the Russians faced stiff resistance almost from the first day, and their assault became embroiled in a series of ambushes instigated by Ukrainian forces armed with NLAW, among other weapons. In response to the failure of the assault and lacking the manpower and supplies focused on Kyiv and Mariupol, the Russians were forced into besieging Kharkiv and bombarding the population with artillery to bring about its surrender. This failed to bring resistance to an end but, by now, the Russians were retreating from Kyiv, which allowed the Ukrainians to launch a series of counterattacks in the Kharkiv area using armour, infantry and artillery to force the retreat of Russian ground forces. As this example illustrates, politics, technology and skill in the form of combined arms warfare allowed the Ukrainians to prevail. Given the high rates of attrition experienced in battle, both sides have had to focus their energies on finding ways to replace their equipment and manpower shortages experienced in this war (Echevarria, 2022).

Conclusion

To what extent does the war in Ukraine validate the assumptions that underlie the vision of future war set out in the United States and other Western versions of MDI? In geopolitical terms, it affirms the idea of the return of great power competition. However, what about the operational assumptions on which the doctrine is based? Do we see this as a grey zone war? If we situate the war within the broader conflict that began in 2014, it is conceivable to see how this conflict chimes with the logic set out in this doctrinal construct. Viewed in this way, what started as a grey zone operation in 2014 then degenerated into a low-level war in those areas of Ukraine contested by Russian minorities living within the recognized boundaries of the Ukrainian state.. The war in 2022 can be seen in one of two ways. The first sees this decision as a consequence of increasing frustration on the part of Russia's leadership. Viewed in this way Russia's political leaders believed their campaign to secure eastern Ukraine was failing and the resort to large scale war in 2022 represented a rejection of a hybrid war strategy, in operation since 2014. Alternatively one can see this as an escalation in violence but it is still part of a hybrid war strategy, which recognizes that conventional military victory is unlikely to happen and that a diplomatic compromise will be required.

Moreover, the action prescribed within the doctrine also seems to coincide with the action taken by NATO. It ensured the Ukrainians were trained and prepared to confront an assault from Russia, which has proved critical in helping them organize their defence. That support was then increased with the direct provision of key military technologies and intelligence to allow the Ukrainians to fight the Russians once the assault began. Where the West erred in not following its doctrine was the failure to keep NATO trainers deployed within Ukraine in the lead-up to war. This might have provided a vital tripwire that stopped Putin from ordering the invasion, but NATO leaders were not prepared to gamble. Instead, they have played a critical role in supplying Ukraine with arms and ensuring it remains in the war. From the perspective of NATO, the aim is not victory but rather, as per the goals specified within MDI, to return to the pre-war status quo. However, in this case, the additional payoff will be a militarily weakened Russia whose reputation for its military prowess and skill will have been tarnished. We seem to have returned to the stereotype of Russian operations relying on brute force to achieve their aims. Russia's lack of skill has also become evident in the failure of its propaganda war. This failed to convince the Ukrainians that Neo-Nazis were leading them or that their fate was better off within a disarmed and neutral Ukrainian state raises interesting questions over the presumption that political warfare waged via social media is guaranteed to succeed. Internationally, world opinion has polarized along predictable lines. However, the brutality of this war in terms of the indiscriminate bombardment of cities and towns, the poor treatment of the civilian population in those areas where the Russians have taken over, and the appalling treatment of Ukrainian prisoners of war has caused considerable reputational damage to Russia's international standing. It is also clear that the West has learned important lessons from what happened in 2014 and has actively contested and denied the Russians access to social media. They have also checked the worst effects of Russia's attempts to destabilize the Ukrainians via the cyber war. The military campaign has proved to be a mix of old and new, and provides hints that support speculative thinking about future war, but both sides lack the kind of technological mass envisaged in MDI. Even so, one striking aspect of the campaign so far has been the rate of consumption of material and how important this has been in shaping the overall tempo and character of the conflict. By April 2022, both sides lacked the vital capabilities and, in the case of the Russians, were forced to withdraw and reorganize and rebuild. This happened again in July as another strategic pause was forced on them after hard fighting to gain control of the Donbas. According to Western intelligence estimates, the Russians lost half the soldiers that formed the initial invasion force into Ukraine in six months of fighting. Despite this cost, there is also little appetite for diplomatic engagement on the part of the Russians, who it seems are determined to achieve a military victory on the ground. Consequently, it is not clear how NATO governments will convince the Russians to accept a return to the status quo; indeed, it is also unclear if the Ukrainians will be willing to accept this as an outcome. This is perhaps one of the oversights of MDI: its failure to recognize that, even if the political objective drives war, it is also subject to the influences exerted by war in terms of the anger and a desire for revenge or justice created in the minds of the belligerents which shape the conflict's political trajectory.

Finally, what does this war reveal about the character of the war–state relationship today, and what can we meaningfully extract from this when thinking about the future? Western concepts of future war give this subject little thought, and it is presumed that the principal function of the state will be to provide the means to wage war and that this will just happen. But what kind of state is required to wage postmodern war? A striking aspect of this war has been the size of forces employed which more closely resemble modern rather than postmodern wars in terms of scale and mass. Second, in this war at least, precision has not diminished the significance of mass as we assumed after the 1991 Gulf War. As Biddle (2022) highlights, Ukrainian mass played a significant role in stymieing Russian offensive action in the first phase of the war. A third feature of this war has been the high rate of attrition of both manpower and equipment, caused in part by the duration of the conflict, but also a consequence of the lethality of the weapons used. This reinforces the importance of deploying and sustaining mass across the technological spectrum. A fourth significant development is the incredibly high rates of consumption of ammunition. According to one report, 'the Russians used more ammunition in two days than the entire British army has in stock' (Zabrodskyi et al, 2022, 55). The fifth is the way in which technology shifted the offence/defence balance in favour of the latter. This shift need not be permanent, but neutralizing a well-established defensive system will require a highly skilled force with access to sophisticated surveillance and targeting systems and the training and weaponry to break through defences. The success of Ukrainian offensive operations in autumn 2022 reinforces this point. However, creating such a capability requires a significant investment in military power, which supports the continued deployment of long-service military that can acquire and maintain the perishable skills necessary to wage a conventional war. This can either be achieved via professional or conscript/militia forces, but the key is to ensure that the right equipment and training are provided.

These observations reinforce the traditional role of the state in the orchestration and conduct of war, and imply the preservation of the status quo. The expansion of war into space and cyber domains reinforces this view and challenges the argument that technology is causing a radical decentralization of power to non-state actors. However, the perspective presented here is firmly based on an analysis of war's operational and tactical levels. In thinking about the future, we must also reflect on the future strategic context and the likely interaction between technology and politics.

Conclusion: Assessing the Impact of the Fourth Industrial Revolution on the Future of War and the State

Introduction

In theory, the future of the war–state relationship seems assured in the Western world. There is a consensus that we have returned to a new era of great power competition, which has been reaffirmed by the recent Russian intervention in Ukraine. Other non-traditional threats are tacitly acknowledged but, as in the past, the driver of strategy and operations remains fixed on the high end of the military spectrum. Within this context, capital-intensive warfare, including nuclear weaponry, and the expansion of warfare into space and the cyber domain will ensure the state remains the principal agent through which increasingly expensive military technologies will be developed, deployed and sustained. This is so even though the cost of these capabilities is increasingly beyond the means of the nation state. The logic of this strategy can be questioned as demonstrated by the use of less expensive dual-use technologies, which in theory, challenge the traditional dominance of platforms like tanks, aircraft and ships. Such a development offers hope to smaller states and violent non-state actors eager to exploit these technological substitutes. However, as illustrated in the last chapter, recent operational experience indicates that specific and local conditions may have allowed cheaper disruptive technologies to prevail in engagements between Russian and Ukrainian forces. As such, it is difficult to extrapolate a meaningful trend from this analysis. In addition, countermeasures are available to challenge these technological disruptors and, for now at least, a conservative orthodoxy continues to shape and drive the war–state relationship. However, while the preservation of the status quo seems plausible and sensible over the coming decade, could this vision of war and the state set out in the previous two chapters be overturned by developments either discounted or not seen in the longer term? In this final chapter, I want to look at how technology in the future will shape and drive war and the state, and their interaction. My argument is that as we move into the fourth industrial revolution technologies, many of which had their genesis as military R&D programmes and were then spun out into the broader economy, are now feeding into politics, economics and society, and this will have significant consequences for the war–state relationship. The potential impact of these changes on war and the state might give us reason to pause and reflect on the vision of war depicted in the latest doctrine manuals in the UK and United States and the extent to which they address the changes discussed here. In doing so, the chapter adopts a longer-term view that looks beyond 2030, where most forecasting stops. This is entirely understandable as looking into the future is a precarious exercise; as Freedman (2017) points out, the future is based on decisions that have yet to be made in circumstances that remain unclear to those looking into a crystal ball (2017, xviii). Tetlock and Gardner (2015) demonstrate the inherent risks of forecasting the future in their study of super forecasting. They retell the story of having examined the predictions made by experts over the previous 20 years and concluding that most longer range forecasts were about as accurate as a dart-throwing chimpanzee. This proposition captured the imagination of the public, and a view emerged that forecasting is a pointless and futile activity (2015, 4–6). Van Creveld offers a good illustration of this mindset in his analysis of this art form (2020b).

In contrast, Tetlock and Gardner (2015) claim that this nihilistic representation of our ability to see into the future presents a distortion of their research. Yes, most forecasts are wrong, but some have proved accurate. As a result, they believe 'it is possible to see into the future, at least in some situations and to some extent' (2015, 6) This is most clearly demonstrated in the mundane and ordinary things we do each day that entail degrees of anticipation of forecasting, for example, leaving for work early to avoid heavy traffic. We cannot predict specific events that might trigger a chain reaction in which the actions of one person then impact millions but, as actuarial science demonstrates, we can at least identify trends based on modelling the available evidence. 'How predictable something is will depend on what we are trying to predict, how far into the future and under what circumstances' (2015, 13).

Most importantly, in defence and security, the only thing worse than attempting to predict the future is not trying. As one study observed, any government seeking to craft its national security strategy must engage with the future 'however unknown and unknowable that future is' (Cornish and Donaldson, 2017, loc 3618). This will allow it to set its strategic priorities and determine the level of resources needed to invest in identified policies and capabilities that best address these threats.

In conducting this activity, it is vitally important to address the biases that shape our preferences regarding how we see the future. As has already been said, this is a particularly acute problem in the military, where visions of future war have frequently demonstrated unconscious bias. Another problem, evident in the current debates about future wars articulated by the military, is that they, the military, often suffer from focusing on technologically sanitized visions of the future battle space, which are entirely devoid of a political context. For example, Cohen (2004, 396) was very critical of the American revolution in military affairs, which caused much excitement in the 1990s because it had nothing to say about geopolitics. This chapter's essential focus is to consider how far the current military vision of future war suffers from such bias. It is also important to realize that the accuracy of forecasting is likely to diminish the wider your time horizon. Simply put, too many unknown unknowns make anticipation difficult, so the vagaries of chaos theory set a clear limit on the art of the possible in terms of forecasting.

One way we can mitigate risk is to follow the advice Mann gave in his book *The Sources of Social Power* (2013). At this time, he acknowledges that no one can accurately predict the future of large-scale power structures. The most one can do is give alternative scenarios of what might happen given different conditions and, in some cases, arrange them in order of probability (2013, 432). Those scenarios are provided by a range of government and academic sources, and there is a general convergence between them. Time and space make it necessary to set out an opposing picture of the future that challenges the new military orthodoxy explored in the previous chapter. I recognize there are several important non-traditional security challenges that might be addressed. However, I have chosen to focus on the potential technological threat because, although merely a means to an end, its long-term impact may well create a very different set of outcomes to those anticipated politically, economically and militarily. This chapter explores the likely impact of ongoing technological change on each of these domains. In so doing, the scenario set out here breaches the commonsense advice offered by Tetlock and Gardner (2015) by looking further into the future over a thirty-year time horizon, which inevitably increases the risk of a wrong prediction. However, the current technological revolution's most significant impact is likely to happen after 2030, so adjusting our own time horizon is essential.

Connecting technological innovation to politics

In the Western world military planning has sought to exploit the fruits of each technological revolution. History tells us that technological superiority can be a vital force multiplier in war, even though that narrative is contested, and the role of technology has done little to secure victory in a succession of recent Western campaigns. Its persistence lies in a combination of factors, including the organizational interests of the services and the political convenience it offers to governments eager not to put their military in harm's way. This assumption continues to drive weapons acquisition across the Western world today and, within this setting, the state plays a vital role in procuring weapons for each service. Nevertheless, have we reached a point where current technological change requires a reappraisal of the utility of technology? Indeed, has the evolution of technology reached a stage where it is now threatening the very thing it was designed to protect – the state? Perhaps not surprisingly, the military vision of future war gives little thought to this question. Instead, it assumes war has a future, and the state will primarily be unaffected by the fourth industrial revolution – at least in the Western world. This assumption has not gone unchallenged (Chin, 2019). If we look specifically at the observations of the US National Intelligence Council in its 2021 report, it noted that: 'Paradoxically, as the world has grown more connected through communications technology, trade, and movement of people, that very connectivity has divided and fragmented people and countries' (National Intelligence Council, 2021, 3).

How then will current technological change shape the state in the future? It is interesting to note that the state is not the only institution struggling to adapt to the current pace of technological change. According to one source, 40 per cent of today's Fortune 500 companies will be gone in ten years, replaced by new firms selling new products and services (Diamandis and Kotler, 2020, 5). This is not to imply that the state will become obsolete, but technological change will possibly produce positive feedback, and this will spill over into politics, economics and society. In the political context, Susskind (2018) believes the connection between technology and politics today lies in its impact on communication and the flow of information that are at the heart of political life. He explains that all political order is built on three elements: coordination, cooperation and control, but all of these rely on a reliable source of information between the ruler and the ruled. 'Language and writing are all forms of information exchange, so too is the concept of bureaucracy, which is largely concerned with the organization of information, its standardization, storage and who has access to it' (Susskind, 2018, 16–17).

A common concern is that these information flows lie mainly in the hands of big tech companies such as Google and Facebook. Like other oligopoly suppliers in history, they are seeking to use this power to achieve a position of market dominance and their wealth is employed to create political power to achieve this goal. It is argued that their needs run contrary to the people's interests. They want a largely unregulated marketplace where they can access all the data we produce as consumers and citizens and sell this information to other companies. At the same time, they do not want the responsibility for policing the internet, but they also hope to resist greater political regulation imposed by the government (Susskind, 2018). The potential hazards of this laissez-faire approach to regulation were demonstrated in the US presidential campaign in 2016 and UK's Brexit campaign in the same year. In both cases a company called Cambridge Analytica was employed to provide data on the profiles of the electorate to help with online campaigns, in the hope of crafting a more effective and targeted election communications strategy. Cambridge Analytica in turn, were able to access the details of over 87 million people registered to Facebook, information that the company sold without reference to the individuals concerned. Although impossible to prove, it is suspected that this data harvesting played an important role in helping Donald Trump become US president in 2016 and shaped the surprise outcome of the UK referendum on whether to stay or leave the EU. Two potential scenarios emerge from this line of thinking. First is what Zuboff (2019) describes as the emergence of surveillance capitalism. This reflects not only a change in the means of production but a coup from above that threatens the operation of any democratic system of governance because of the high concentration of wealth it produces (Zuboff, 2019, Chapter 18). In her view: 'A new technological aristocracy is emerging who are accountable to no one. As such, it imposes a set of social relations resembling premodern absolutist authority' (2019, loc 217). This damning critique of high-tech capitalism represents a broader attack on the political order of the West. Fukuyama (2015) refers to three component parts of a political order when using this term. These are the bureaucracy, the law and the creation of accountability between the ruled and the rulers (Fukuyama, 2015). However, these are also vital components of the state in that they provide the means to achieve a consensus on national goals such as security. It is also essential to recognize that political order will vary from one state to another, and it is interesting to see how different the relationship is between the Chinese state and their big tech companies, where the balance of power lies squarely with the former (Strittmatter, 2019). In this case, private companies play a direct role in securing the state by allowing the government to see into every corner of people's lives. This has also happened in the West, for example, during the war on terror, a significant effort was made to use the likes of Google to enhance the surveillance of the state (Zuboff, 2019, loc 219), but there are also examples of where companies in the West, such as Apple, have refused to cooperate with the state to allow it to monitor the users of its products and services. So, in political terms, the fourth industrial revolution is causing a concentration of political power in the hands of people who do not govern directly and are not necessarily interested in the general welfare of the community that makes up the state. This raises an important question of the legitimacy of governance within the state.

Technological change also leads to distortion of the truth within the political order. Currently, the internet of things has over ten billion devices attached to it, and this is expected to grow to over 64 billion by 2025 and possibly trillions as we move into the 2040s (National Intelligence Council, 2021, 2). This explosion in connectivity creates opportunities but also leads to a bounded rationality. As P.W. Singer (2018) noted, one of the strange things about the internet and the increased flow of information is that algorithms identify the things we are most interested in and select from this vast information stream. As a result, our views and beliefs are reinforced rather than challenged, which can lead to radicalization as people become increasingly entrenched in their perspectives which can lead to the polarization of political opinion within society (Singer, 2018, 151–4). This phenomenon raises the profound question of what exactly constitutes the truth. Each side can support their argument with data and facts, which creates certainty in the views people hold and righteous indignation when they are challenged. Most importantly, our sense of what constitutes truth comes to be shaped by algorithms, which are created and owned by companies. Such a development poses a significant problem in democratic states which rest on the emergence of a consensus on how to govern in the best interests of the people. In this case, technology amplifies political ills that already exist within Western societies, leading to the polarization of opinions and the adoption of extreme positions on specific issues, for example, immigration. Radicalization via the internet is a well-known phenomenon, most usually associated with Islamist terrorism and the war on terror. It is widening, and I am referring principally here to the emergence of conspiracy theories employed by right-wing extremists within Western states, in which political leaders such as Donald Trump are portrayed as the people's saviour. Only he is taking on a political establishment that is believed to be corrupt. Perhaps the most bizarre illustration of this idea that Trump is a hero was his supposed role in destroying a global paedophile ring which included many of the world's most notable political leaders, including the British royal family. A more recent example of how technology has made it ever more challenging to identify the truth is Trump supporters' storming of the US Congress on 6 January 2022. This event was justified because people were convinced their man had won the presidential election, even though every legal challenge to the outcome of the 2020 presidential election launched by Trump's lawyers was rejected by state judiciaries across the United States. Trump is also accused of playing an instrumental role in causing rioting in the capital through his use of Tweets to his supporters before Joe Biden's inauguration. This reveals how political leaders can exploit technology to mobilize and bypass the laws and regulations on the transfer of political power within the state. Technology is not just changing politics. It is claimed it is also on the verge of precipitating an economic revolution which will feed back into politics and society, which takes us into the next projection of how technology will impact on the state in the future.

Technology and economics

The biggest worry here is what the fourth industrial revolution and, in particular, the growth of machine learning and AI will do to the jobs market as these technologies are used more and more to replace people in the workspace. The business world is years ahead of the military in terms of its use of machine intelligence. As McAfee and Brynjolfson (2014) explain, innovation was hugely advantageous in those occupations that relied on physical labour, allowing new forms of economic activity and employment based on human cognitive abilities to develop. However, this cognitive comparative advantage is now under threat, as computer algorithms have reached a point where they can outperform humans in many jobs (Harari, 2017, 363).

Before investigating the likely effect this revolution will have in the West, it is crucial to acknowledge this is not going to be a local event confined to the West, but will also have a severe impact on the developing world. The introduction of AI removes the comparative advantage of these states in the form of cheap labour, which allowed the developing world to compete and relocate global manufacturing into poorer countries. The growth in AI and automation reduces the importance of cheap labour. According to Stiglitz and Korinek (2021), the effects of this revolution will extend beyond manufacturing in the developing world into the service sector, where the use of algorithms will improve labour efficiency in the many call centres that relocated from the West to cheaper alternatives such as India, and AI could precipitate the return of these jobs to the West. Even developing countries that rely primarily on the export of commodities will see demand for their resources reduce as more efficient forms of consumption are mapped out by AI (2021, 6).

The precise impact AI will have on Western economies in the future is contested. Of specific importance is the question of how many people might lose their jobs and what the consequences of this will be for national economies, politics and, more generally, society. As Stiglitz and Korinek (2021) observe, there is a general assumption that technological advance is a good thing and will benefit all. However, there is no economic theory that supports this optimistic assumption. In fact, the opposite is the case and economic theory clarifies that technological change can lead to dislocation and loss (2021, 2). In the case of the West, attempts have been made to estimate the likely cost of this latest tech revolution, but the conclusions paint a rather depressing picture.

A Price Waterhouse Cooper (PWC, 2018) report predicted that 38 per cent of all jobs in the United States will be at high risk of automation by the early 2030s. Most of these are routine occupations such as forklift drivers, factory workers and cashiers in retail and other service industries. This depressing analysis is supported by the Bank of England's estimate that up to 15 million jobs are at risk in the UK from increasingly sophisticated robots, and that their loss will widen the gap between rich and poor (Elliot, 2015).

Most worrying is that, in the short term, the jobs most at risk are low-paid and low-skilled occupations, which are precisely the jobs the UK and US economies have been so successful in generating to create record levels of employment since the financial crash in 2008. From the perspective of firms, this transition makes economic sense. The dramatic reduction has matched the enhanced performance of robots in cost. Today a Danish company, Universal Robots, sells its robots for US$23,000, equating to a factory worker's global annual wage. However, in the case of the robot, this is a one-off payment, and the robot does not tire or suffer from boredom and can work without a break for many hours. Not surprisingly, Ford, General Motors and Tesla are fully automating their plants. Foxconn, which manufactures the iPhone, is also automating its production processes. Amazon not only relies increasingly on robots in its warehouses but is also pushing forward with the use of drones in the delivery of goods ordered online to its customers (Diamandis and Kotler, 2020, 47).

As in the past, those most affected by this change will be the economically least powerful sectors of society – the old and unskilled and unorganized labour. Until now, the managerial and professional classes have been able to use their economic and political positions to protect themselves from the worst effects of such crises (Gurr, 2015, 58). The big difference about this revolution is that AI is threatening traditional professional middle-class occupations. Any job that can be done using pattern-searching algorithms will be vulnerable. This includes banking and finance, the law and even education. Daniela Russ (2015, 2–6) argued that humans need the personal touch in their day-to-day lives and that humans are therefore guaranteed to have a place in the job market.

Harari challenges this view, and claims that machines can mimic empathy by monitoring blood pressure and other physical indicators in interactions between AI and humans (Harari, 2017, 370). A report by the *Wall Street Journal* (2018) supports this view. In their investigation of the use of AI in the provision of psychological therapy, they found that people preferred the treatment offered by the AI precisely because it was a machine, so they did not feel judged. The system can also be configured to fit people's preferences, creating a 3D computer-generated image that is comforting and reassuring. A significant limitation of AI and machine technology is that currently, it cannot replicate the dexterity of humans in handling delicate objects, leaving a role for humans in the workplace. However, scientists in California are looking at the use of AI and machine technology as a way of addressing the acute labour shortages experienced in the fruit-picking industry; this includes the development of machines capable of deciding which fruit is ripe for picking and doing so in a way that does not damage the product during picking, processing or distribution. Given these developments, Harari's prediction for humans in the workplace is bleak. 'In the twenty-first century, we might witness the creation of a massive new unworking class: people devoid of any economic, political or even artistic value, who contribute nothing to the prosperity, power and glory of society' (Harari, 2017, 379). The mass unemployment generated would be on an unprecedented scale and likely to precipitate instability and possibly violence (Drum, 2018, 47).

Further evidence to support the depressing scenario depicted here is provided by the former head of Google China, Dr Kai-Fu Lee, a man with decades of experience in the world of AI. In his view, AI 'will wipe out billions of jobs up and down the economic ladder' (Kai-Fu Lee, 2018, 19).

A typical counter to this view is that AI will lead to the creation of new jobs and new careers, but, as Tegmark (2016) explains, the evidence does not support this claim. Looking back over the last century, what is clear is that 'the vast majority of today's occupations predate the computer revolution. Most importantly, the new jobs created by computers did not generate a massive number of jobs' (Tegmark, 2015, loc 2066).

This view of economic despair has been challenged. A study by McKinsey Global Research found that the internet created 2.6 jobs for each one extinguished by this technology (Manyika et al, 2020). Diamandis and Kotler (2020) believe this change pattern will be replicated in many industries. Based on their internet experience, they claim that more wealth could be created over the next ten years than the previous century (2020, 22–23). Any assessment of how technology will lead to mass unemployment must also break work activities down so we can see more precisely how technology and humans might interact within the same space. While certain parts of a job could be done better by a machine, there are still significant areas of activity that rely heavily on human interaction, which cannot be replicated easily by a machine. Viewed in this way, perhaps AI will not destroy jobs but change them so that workers spend more time on the social interaction aspects of their work. So, for example, doctors will not be replaced by robots, but more of what they will do will be automated, allowing them to focus more on working with people (Baxter and de Salaberry, 2020, 162).

Critics also point out that history shows that previous industrial revolutions destroy jobs and created them. A report by the World Economic Forum (2020, 163) forecasts that by 2025 machine technology could eliminate 75 million jobs but further into the future it might compensate for this loss by creating 133 million new positions. There are times when time lags between innovation and wealth creation lead to people becoming worse off in the short term. Are we experiencing another of those moments now?

The problem is that these estimates are based on our experience of past industrial revolutions, which have created new jobs to replace old ones. In this case, the revolution is taking place mainly in the cognitive domain, and it is not clear that this will create new jobs. One career that has taken off due to current technological change is the data scientist, whose task is to provide greater insight and understanding, a skill that is becoming increasingly important. Does this mean we are on a downward trajectory in economic terms? According to Baxter and de Salaberry (2020), the answer to this question is not clear-cut. They note that the first industrial revolution coincided with a period of falling median wages and weak economic growth. In the case of the second industrial revolution, the world still endured economic disaster in the form of the Great Depression in the 1930s, the rise of extremism and world war (2020, 164). However, this offers some cause for hope in terms of the latest revolution leading to the creation of new jobs. But, in the short term, at least, there is 'enormous doubt concerning the job prospects of swathes of people who work in areas subject to wholesale automation. Will they acquire the skills to operate in the newly created employment roles? An entire generation, or perhaps two generations, might be lost as they will not be sufficiently skilled to meet the requirements of the new age. According to research from Ball State University, technology/automation was the leading cause of the loss of well-paid jobs in US manufacturing (Hicks and Devaraj, 2015).

The problem is not just about jobs; the inequality created by the latest revolution is as important. Although salaries are greater than 40 years ago in the United States, their purchasing power has not increased. Income inequality today is perhaps higher than in any other era. This produces a range of social and political problems, leading to technological stagnation. For innovation to take hold requires that it exist in a setting where the fruits of this revolution can be sold to a mass market, but the prevalence of inequality undermines this environmental setting, which undermines the idea of a mass market. This is the main way in which innovation takes place. The key to innovation over the past 200 years has been the creation of such mass markets. Within this context, globalization has played a vital role in opening markets. For this to work, people need to be able to afford to buy the technology. Poverty in the developing world slows technological advances. So for example, the rise of China has helped fund the development of AI (Iansiti and Lakhani 2020, 166).

Technology is also changing markets, and we are moving away from an era of mass consumption as technology is used to increase the life span of products, reducing the speed at which they are replaced. In addition, streaming services and vast electronic storage in the form of the cloud mean we no longer own things like music, DVDs, books or photos but keep them in the cloud. The introduction of self-driving cars was also expected to precipitate a fall in ownership Iansiti and Lakhani (2020, 248).

Viewed from the perspective of today the existing data indicates we have a shortage rather than a glut of workers in the employment market. Currently, unemployment is at a 50-year low in the United States and the UK. Based on this fact, one journalist observed that all the predictions about mass unemployment caused by robots taking over all our jobs are wrong: 'very little decisive evidence has emerged to support the broad thesis that robots are bad news for jobs. On the contrary, much of the world is experiencing an acute labour shortage' (Samuel, 2022).

How then do we explain employment trends that challenge mass unemployment predictions? Are these forecasts wrong? Ford (2022) has examined this problem and concluded that this surge in employment could be explained by reference to short-term factors that will dissipate as we move into the future. One illustration of this was the extended use of government financial support, which allowed millions of workers in the UK to collect 80 per cent of their salaries during the severest episodes of the COVID-19 pandemic when restrictions on movement prevented many people from going to work. This protected jobs and provided an incentive for employers to retain staff. It was feared that once this scheme ended, mass unemployment would follow, but this did not happen. In the case of the UK, it did not because the UK had departed from the European Union. This event resulted in 200,000 EU citizens leaving the UK and returning home. The inevitable consequence was a shortfall in the workers required in the agriculture, hospitality and logistics sectors. Ford believes the data showing full employment is also misleading and reveals a deeper underlying problem in the job market in that it does not consider the increasing numbers of people who have simply opted out and are no longer registered on the system as ready and available to work. This trend was most apparent among male workers who struggled to find anything more than low-paid work, and as a result, there has been a 10 per cent decline in the number of men in the United States actively searching for a job. Post-2000, increasing numbers of women also joined this group by voluntarily withdrawing their labour from the marketplace (Ford, 2022, 140).

Trends have created a powerful incentive for businesses to invest in automation to address the shortage of workers. In the UK, the absence of seasonal labour from Europe has increased interest in developing agricultural robots capable of removing weeds without damaging crops. Another company, funded by venture funding from the UK government, built a robot capable of harvesting fragile fruits and vegetables in greenhouses and uses AI to select the ripest produce. Machines are also being used in restaurants to cook burgers and fries. This also reduces the chance of food contamination, helpful in a post-pandemic world. McDonald's is currently experimenting with AI to process food orders in some of their drive-thru restaurants in Chicago. While it is unlikely these technologies will provide an immediate solution to the ongoing shortage of workers, there is little doubt that it will impact the jobs market.

Amazon uses over 200,000 robots in its distribution centres worldwide; Ocado employs over 1,000 robots at a single distribution centre in Andover, Hampshire. Labour continues to be employed in these warehouses because robots are not yet capable of picking and stowing objects, but this is likely to change in the future as both Amazon and Ocado are investing in the development of robots with greater dexterity. Jeff Bezos predicted in 2019 that the problem of robotic grasping would be solved within a decade. As a result, many people working in these distribution centres are likely to become unemployed in the near future. Once this is achieved, it is only a matter of time before it is deployed in supermarkets and in restaurants.

Ford is also sceptical that the technological fruits of the fourth industrial revolution will create new jobs on the scale needed to replace those lost. In the past, automation of one sector of the economy has frequently led to job losses, but it was possible to absorb the unemployed in other sectors of the economy that were expanding. So, for example, in the 1970s, it was believed that automation in UK factories would result in the reduction of workers required to operate production lines, but it was believed that those who lost their jobs would move into the expanding services sector of the economy in finance and retail. Today 80 per cent of workers are employed in this sector, which forms the most significant part of the UK and US economies. This is unlikely to happen in the future because AI is what Ford terms a general-purpose technology. This means it will apply to all sectors of the economy – even low-tech sectors of activity such as agriculture. Any new industries will also be defined mainly by the impact of AI. 'In other words, it seems unlikely an entirely new sector with tens of millions of jobs will somehow materialize to absorb all the workers displaced by automation in existing industries' (Ford, 2022, 165).

Second, many more jobs are vulnerable to the AI revolution. Ford estimates that half the workforce is engaged in occupations that are primarily routine and predictable and are therefore vulnerable to automation generated by AI. Where will these people go, and where will they find other routine-based jobs in the future? Given the decline in the numbers of such positions, people will face a considerable challenge in transitioning into work that is not routine or predictable; but will there be enough of these jobs for all, and will people be able to make the transition? (Ford, 2022, 166).

Of particular importance here is confirmation that this revolution will have a significant impact on white-collar jobs. Any job involving the routine manipulation of information is likely to lend itself to automation. This is already happening: in some of the world's largest media organizations, AI is being used to generate news articles. Algorithms are also being used to analyse legal contracts and predict the outcome of litigation. The critical point is that knowledge work will be more accessible and less expensive to automate than lower-paid work requiring physical manipulation. In this knowledge setting, there is no need for an expensive mechanical robot and no need to resolve the problem of replicating human dexterity or mobility (Ford, 2022, 181).

Most importantly, the meteoric increase in the capability of AI in the future means that the dividing line between routine jobs, which are vulnerable to automation, and those that are safe will change as AI becomes increasingly capable. Ford speculates that, as we move to a world where programming of new algorithms is done increasingly by machines not encumbered by the restrictions imposed by a human programme, AI creating code for knowledge-based tasks will become increasingly common. We can already see how software automation is encroaching on a broader range of activities, including those requiring judgement. Bloomberg uses systems to automate essential journalism by analysing a stream of data, identifying the story and crafting a narrative text. The release of ChatGPT in November 2022 and its update, Chat GPT4 in early 2023, is demonstrating how AI might change virtually every aspect of our lives now and in the future. Developed by OpenAI, a research organization in California, ChatGPT is more than an online encyclopedia. It can write essays and passable poetry. It understands images, apparently it can even give you a recipe based on a photo of what's in your fridge. It can tell jokes on any subject provided. On a practical level companies are already using this capability to help in the construction of technical and business reports and it is feared it will lead to the death of essays in higher education as university students use ChatGPT to write their research papers.

Analytical jobs in banking will also prove to be vulnerable, and it is predicted that automation in the US banking system will result in 200,000 job losses (Ford, 2022, 186). Call centres are another area where technical improvements in language will result in greater automation of this service and the retention of a tiny human staff to deal with the most complex problems. Several companies already offer AI-powered chatbots to automate customer services. Machine learning will ensure that these systems become more innovative and will be able to deal with an increasing array of customer problems. We are now reaching a point where even computer programming is becoming automated; in essence machines are designing software for machines. Both Facebook and DARPA are investing in this capability. Ford concludes:

> As technology begins to encroach on even these more educated and highly paid workers, inequality is likely to become ever more top-heavy,

with a tiny elite that own vast amounts of capital pulling away from everyone else. As better paid workers are increasingly impacted, this will further undermine consumer spending and the potential for robust economic growth. (Ford, 2022)

He continues: 'If the 4th industrial revolution is likely to lead to a world of mass unemployment, how will this feedback into politics and how will the political discourse created then feed into the process of conflict and violence? (Ford, 2022, 188).

The political impact of technological change

Although depressing, the scenario depicted above does not mean we are condemned to what Martin Wolf (Wolf, 2015) describes as a kind of 'technological feudalism'.

The state has dealt with economic shocks before, as demonstrated by the substantial fiscal stimulus injected into Western economies during the COVID-19 pandemic. Moreover, as Gurr (2015, 59) points out, past economic crises have provided political incentives for social reforms: for example, the New Deal in the United States represented a revolutionary change in how the central government sought to manage the economy.

According to Wolf (2015), three factors might determine how well the state deals with these challenges: first, the speed and severity of the transformation we are about to experience; second, whether the problem is temporary or likely to endure; and third, whether the resources are available to the state to mitigate the worst effects of these changes. In the past, Western governments have deployed a range of policies to deal with recessions or, as in the 1970s, scarcity of resources such as oil. However, these macroeconomic policy responses operated on the assumption that such crises were temporary and that economic growth would resume and normality be restored quickly if the proper measures were in place. In contrast, the AI revolution is happening rapidly, and will be an enduring feature of economic and political life. In Wolf's view, this latest revolution will require a radical change in our attitude toward work and leisure, with an emphasis on the latter. He also believes we will need to redistribute wealth on a large scale. In the absence of work, the government might resort to providing a basic income for every adult, together with funds for education and training. The revenue to fund such a scheme could come from tax increases on pollution and other socially negative behaviours. In addition, intellectual property, which will become an essential source of wealth, could also be taxed (Wolf, 2015, 22).

However, introducing these measures will not necessarily prevent a rise in political protest and or organized violence. As Gurr (2015, 16) explains, recourse to political violence is caused primarily not by poverty but by relative deprivation. This is defined as 'actors' perception of discrepancy between their value expectations and their environment's apparent value capabilities'. It reflects the difference between what people believe they are legitimately entitled to and what they achieve, perceptions of which have become acute in the smartphone age. Relative deprivation applies both to the individual and the group. In this light, the bright, shiny new world created by AI provides a potentially rich environment for relative deprivation – particularly if large swathes of the middle classes are frustrated in their ambitions and suffer a loss of status as a socioeconomic group. More worrying is that this technological and economic revolution will coincide with the global deterioration of the environment, which also challenges the state.

Within this scenario, states in the Western world will struggle just as much as states in the developing world. If the legitimacy of the state is measured in terms of its capacity to effectively administer a territory under its control, then the political context set out here poses a significant threat to this institution. The extraction of resources through taxation will prove extremely difficult as the tax base shrinks. This will affect the ability of the state to provide the public goods the population expects and requires. A weaker state, which lacks the resources and capacity to sustain the population, will also lack legitimacy, reinforcing the political problems caused by technology described above; this could cause the social contract to break down and result in widespread protest and possibly violence.

However, economic factors alone do not explain protest and rebellion, and within the context of democratic states, there are usually other more peaceful avenues through which a grievance can be addressed. For example, the right to strike is usually enshrined in law in Western states. Such action can also have a strong political impact; in the past, UK organized strike action literally humbled and eventually caused governments to fall in 1973 and 1979. However, the world facing organized labour today is very different to that which prevailed in the 1970s, and it is questionable whether striking is either an option for individuals who work in a very different social and political setting to their predecessors. In part this is because governments eager to avoid confrontation with the organized unions introduced legislation that constrains their power to strike. However, more important from the perspective of this study is the role played by technology in precipitating the devaluation of labour and thereby reducing its bargaining power in terms of securing higher wages.

Ford (2022) explains that this has happened because of the decoupling of productivity increases, which has not been reflected in wage growth and has led to salaries stagnating or falling in real terms. Productivity is calculated by taking the total economic output and dividing it by the number of labour hours required to generate output. It is a vital economic indicator and is fundamental in facilitating economic growth. It should also mean that worker salaries increase in real terms, which helps raise living standards for ordinary people. However, since at least the 1970s, wage increases have failed to match rises in productivity. As a result, nearly all the gains from technological progress and improving productivity are now being captured by a small group of wealthy people who account for about 12 per cent of the workforce (2022, 171). Again, Ford believes this divergence between productivity and income for the vast majority has been caused by the introduction of machine technology. Once these systems became autonomous, they ceased to complement labour that ensured rising productivity was shared, and came to be a replacement for skilled and semi-skilled labour. As a result, there were fewer skilled jobs and jobs in general, weakening the labour bargaining power. Lower wages then led to an increase in income inequality. So, there are jobs available, but they tend to be low-wage jobs in retail and food preparation. Equally important, these are casual and offer no security, a development captured in the idea of the gig economy. In 2019 it was estimated that 44 per cent of the US workforce was engaged in low-wage jobs (Ford, 2022, 173). As important here is the impact of globalization as jobs in offices and factories have been exported overseas.

New high-skilled jobs have been created, but these are not accessible to the 75 per cent of Americans who do not have a four-year college degree. However, even those with a degree struggle to secure graduate-level jobs. In the United States, over 40 per cent of graduates are engaged in jobs that do not require a college degree. Unemployment among graduates is also double the national average. This means the economy is simply not producing enough graduate-level jobs (Ford, 2022, 175). Given how easy it is to replace labour with machinery, a trend that will become increasingly pronounced in the future, it seems unlikely that organized labour will have the bargaining power of unions in the past and consequently is more likely to fail than succeed. This then leads to more direct involvement in the political process. Beissinger (2022, 9)has observed that social revolutions in the 20th century were dominated by 'grievances over poverty, redistribution of wealth, and land inequality'. In contrast, today's political action is mainly in response to government repression, corruption and misrule. Beissinger also acknowledges that economics remains an essential motivation for protest or revolution, and he sees a strong connection between the rise of economic grievance and the emergence of revolution. Usually, this manifests itself in an urban setting, where most people now live, and he sees the middle classes within this constricted environment as the key agents of change. As he explains, modern urban living 'engenders specific desires and demands (like paid jobs, regular pay, particular norms of consumption); it simultaneously inculcates among urbanites a set of entitlements and rights to have optimum urban services. When states cannot provide these services, this can provoke outrage and action' (2022, 9–11). Sometimes the result is what he terms urban civic action, which amounts to mass protest, which is usually focused on the removal of the government. This is presented as a modern version of political revolution and seeks to employ technology in the form of smartphones and social media platforms to facilitate and coordinate the mass mobilization of the people. This argument becomes more compelling if we superimpose on it the predicted jobs crisis coming towards us. In the past, mass protest relied on a dedicated party structure based around committees whose function it was to direct the membership, but today, as Beissinger explains, there is no formal organization or structure; in its place is a rapidly formed coalition brought together through the power of social media and usually united in their opposition to the government (2022, 13). Protest of this kind, for example, the Arab Spring in 2010, produced mixed results, and there is no guarantee that the protesters will succeed. It is essential to recognize that if all else fails, the state has recourse to the use of brute force, even though this might damage its legitimacy and reputation at home and abroad. Protest against the results of the Iranian presidential election in 2008 and, more recently, Syria's response to the Arab Spring provides a salutary reminder that the state retains a considerable advantage in terms of its monopoly on the use of force. It is also important to remember that technology is a double-edged sword and provides the state with several important advantages. Stephen Graham (2011, loc 2011) notes that a significant trend in the war on terror was the blurring between civilian and military applications of technologies dealing with control, surveillance, communications, simulation and targeting.

The capability to exercise control via that which are intended to provide a service, such as parking and congestion charging, has dramatically increased the opportunities to conduct electronic surveillance for a host of other purposes. As has already been said, technological change also creates opportunities for governments to centralize political power, as China is currently doing. Indeed, the pursuit of control is one of the main reasons China invests so heavily in AI.

War, technology and the state

If strikes and protests fail to pressure the government to address the needs of the people, then the use of violence might become a possible option. However, it is essential to recognize that this constitutes a big decision for the individual. Preparing to strike and protest does not pose a moral dilemma for most people. However, to use violence to achieve one's goals entails putting oneself and others at risk, which raises critical ethical questions. The literature on radicalization and terrorism seeks to address this question and ask why radicalized people choose to engage in violence while others do not but, so far, it has not provided a definitive answer to this question. Instead, a range of factors are cited, focusing on the individual and how badly they are affected by the current crisis they face – is it possible to justify targeting members of a government implementing policies that are causing direct harm to strangers? Of equal importance is the environmental setting in which the proposed campaign is to be implemented – the execution of violence is easier in certain environments, but the fundamental question here is can violence facilitate success? Assuming action is to take place within the city, which seems a reasonable proposition given the distribution of the population in the West, the advantages held by the state's security apparatus are significant. As such, the resort to violence is not straightforward. This was why Fidel Castro declared that the city was the graveyard of the revolutionary. However, technological innovation might change the risk calculation to the extent that people become increasingly willing to use violence.

In sum, the fourth industrial revolution is leading to the creation of capabilities that largely reinforce the state's power through surveillance technologies empowered by AI. However, technology also provides the means for individuals or groups to embark on various forms of direct action and even violence. In this future version, the most fundamental aspect of the technology–war interaction will be the challenge to the state's retention of the monopoly of violence. Projections about the end of the state's monopoly on the use of force have been made before. The most notable of these commentaries is Martin van Creveld. Over the past three decades, he has consistently argued that the demise of the state is on the horizon. Within this context, he referred to wider systemic forces, such as globalization, the rise of regional governance, economic and demographic problems and most recently, the West's cultural decline, all of which were combining to create a challenge for the West and indeed for governance in both the Western and non-Western world (van Creveld, 1991, 2016). As he observed:

> In the future, war will not be waged by armies, but by groups whom we today call terrorists, guerillas, bandits, and robbers ... Their organisations are likely to be constructed on charismatic lines rather than institutional ones and to be motivated less by professionalism than by fanatical, ideologically based, loyalties. (van Creveld, 1991, 197)

The passage of time has not been kind to this thesis. Thirty years on, the state continues to persist, but the power of globalization and regionalism is receding. As has been said, there is also a renewed focus on the importance of the external rather than the internal threat to the nation state posed by other nation states. However, it is possible that, in the future, different internal and external forces might combine to generate a new crisis in the state and the creation of conditions which undermine the ability of the state to make war along the lines implied by van Creveld (1991, 1999) and others. One of these factors will potentially be technological innovation, a process that helped cement the war–state relationship and might soon provide the key to unlocking our Weberian conception of the state.

In the case of the internal driver of this crisis, the fundamental problem lies in the diffusion of technologies that provide cost-effective alternatives to existing military capabilities. This is not a new argument, and we can go back to 1999, and the book *Unrestricted Warfare*, which was authored by two colonels in the Chinese People's Liberation Army, Qiao Lang and Wang Xiangsui. This study was conceived mainly within the context of a future war between the United States and China, and so their thinking was developed within the setting of a state-based conflict. However, their central thesis is relevant here because they believed the world was living in an unprecedented age in terms of the speed and breadth of technological innovation. They argued that so many essential technologies are emerging that it is difficult to predict how these will combine or what the effect of these combinations will be in military and political terms. Developments in biotechnology, materials technology, nanotechnology and the information revolution are creating new opportunities and ways of attacking other states. An important observation made in *Unrestricted Warfare* is that new technologies, which could be used as weapons, are increasingly part of our everyday day-to-day lives (2015, loc 273).

In sum, the colonels identified a range of non-military means that are technically outside the state's control and might allow a weaker actor to fight and defeat their more powerful adversary. The 20 years that have passed since the publication of *Unrestricted Warfare* have demonstrated the prescience of the authors in respect of what are deemed to be new types of conflict today. For example, what they called 'super terrorism war' seemed to come to fruition on 9/11. We can see how state and non-state actors have exploited emerging technologies that challenge the nation states. Of great importance is how groups such as Islamic State in Iraq and Syria (ISIS) and revisionist powers such as Russia have weaponized social media in their efforts to weaken those who oppose them. ISIS, indeed, claimed that media weapons could be more potent than atomic bombs (Singer and Brooking, 2018, 151–4).

An excellent example of this trend can be seen in synthetic biology, a new field combining computing and biology to 'design and engineer new biological parts, devices and systems and redesign existing ones for other purposes' (Noble, 2013, 47). In 2003, the Human Genome Project completed the first whole sequencing of human DNA. The successful completion of this project took ten years and resulted from work done in over 160 laboratories, involving several thousand scientists and costing several billion dollars. It is now possible to buy a DNA sequencing device for several thousand dollars and sequence a person's genome in less than 24 hours. So steeply have sequencing costs fallen that the industry is no longer profitable in the developed world and is now primarily conducted within China. The potential of synthetic biology to reshape our world is so great that it is claimed it will have an impact as great as the information technology revolution (Pavel and Venkatram, 2021). However, synthetic biology has a darker side.

An example of the potential threat posed by this new science emerged in 2005. Scientists, worried about the possibility of another flu pandemic, recreated the Spanish flu virus, which, during 1918–19 killed 50 million people. In 2011, scientists also employed these techniques to manipulate the H5N1 bird flu virus and create a variation that could be spread from the avian to the human species. Concerns were raised over the necessity of work of this kind and the possible access to highly sensitive information which was published on these developments. At the time it was feared a violent non-state actor might be able to exploit this knowledge and develop their own bioweapon. Past precedent supported the view that some groups, for example, the Japanese cult, Aum Shinrikyo, were strongly attracted to the use of such a weapon to fulfil their millenarian dreams. Islamist terrorist groups such as al Qaeda also demonstrated a strong interest in developing this capability (Garrett, 2013, 28–46).

While it is difficult to speculate what ideology will emerge in the future and the new forms of terrorism which might arise from it, there is a concern that developments in synthetic biology have reduced both cost and the technical barriers to entry into this domain to a point where it so sufficiently low that it can be exploited for nefarious purposes by individuals or groups (Pavel and Venkatram, 2021). This general anxiety has been compounded by the legacy of COVID-19. This global pandemic revealed how vulnerable modern societies are to disease and how enormously costly it proved for governments to contain its spread (Kurth Cronin, 2022; Pavel and Venkatram, 2021).

Precisely the same fears have been expressed about the cyber domain. According to one Israeli general, 'cyber power gives the little guys the ability that used to be confined to superpowers' (Naim, 2014, loc 2571). 3D printing also reinforces the diffusion of power, in theory, it is possible to build a handgun or even an assault rifle with this technology.

However, before concluding that the state is about to wither away, we must remember that these technologies are still maturing. Therefore, whether or not advances in the cyber domain will undermine or reinforce the state's power remains a contested point. As Betz (2012) pointed out, launching a successful attack against another state via this medium can be very costly. The Stuxnet computer virus, used to attack Iran's nuclear programme, was a very sophisticated piece of software developed by a dedicated team of specialists over a long period. The successful insertion of this virus also required high-grade intelligence on the Iranian nuclear programme. Consequently, the success of a cyber-attack depends on a combination of capabilities, not just the development of a virus, and at the moment, this puts the state at a considerable advantage (Betz, 2012, 695). However, this attack was designed to hit a hardened facility and as such was heavily protected from possible remote attack in the physical or virtual domain. Our ever increasing reliance on information technology and the predicted growth in connectivity in technological terms politically, economically and socially will increase our vulnerability to even relatively simple attacks. In wars between states, it is clear that cyber-attacks will conducted across society, business and government in an effort to disrupt the state's efforts to prevail in a future war. The significance of cyber war can be seen from the fact that Western militaries have made this into a separate domain of war which sits alongside land, air, sea and space (see Chapter 5). The war in Ukraine demonstrates both countries have invested heavily in waging war in this area. Non-state actors, seeking to cause harm but lacking resources to contemplate large scale sophisticated attacks will still have the means to cause harm, albeit at a lower level in the form of defacing websites, overloading computer systems through distributed denial of service attacks. Ransomware, used largely for economic gain, has proved to be potent instrument for extorting money from individuals, but more significantly large business organizations; for example, the Wannacry attack in 2017 and Colonial Pipeline attack in 2021. More worrying is the emergence of 'Weaponized Operational Technology Environments'. This refers the possibility of hackers shutting down parts of a network to stop the provision of a critical service, for example closing down a hospital's information technology systems, which would lead to the death of patients (Haughey, 2021).

A similar point can be made in the case of 3D printing: you need to do more than just download the code to print the weapon. You also need access to complicated and expensive computer-aided design software and a high-quality metal 3D printer capable of using steel, aluminium or nickel. Such a machine costs over US$100,000, which is nearly 60 times the price of a standard 3D printer which uses plastic. The latter has been used to print plastic guns, but these proved unreliable and are likely to explode in the user's hand (Tynan, 2018). However, ongoing technical improvements in the area of 3D printing have resulted in the development of less expensive and more reliable weaponry. As Banerjee (2021) points out, in 2019, two people were shot in Halle, Germany by a person using a homemade weapon, based on a blueprint downloaded from the internet which was used to complete the construction of the weapon. In 2021, Spanish police closed down an illegal workshop in the Canary Islands which was producing 3D printed weapons and in the same year, three people were arrested in the UK as part of an investigation into right-wing extremism, and all were found to be in possession of components of 3D printed weapons. Moreover, even if the cost of 3D printing remains prohibitive, it might still prove profitable to print certain parts that can be used in a weapon. For example, 3D printing can be used to make a lot more than just firearms. It is possible states and even non-state actors might use this capability to evade measures intended to prevent the proliferation of equipment that could be employed in the construction of a nuclear device. 3D printing was also used by Raytheon to manufacture most parts of one of its guided missiles (Banerjee, 2021).

On presentism and technological determinism

Current technological change appears to create two distinct trajectories that will impact the relationship between war and the state. The first reinforces the status quo, and the second undermines it. The first represents the crafting of technology and the associated doctrine and organization in a manner that reinforces the importance of acquiring 'beyond state of the art weaponry'. The need for such technology is justified because potential enemies are also investing heavily in weapons to enhance their own security, triggering a technologically driven arms race. The focus on this kind of technological development makes sense when viewed within the context of great power competition. This framework largely preserves the state's role in providing the financial means to fund research undertaken by the private sector and the auditing of their work by various defence, scientific, technical and accounting organizations within the state. This does not preclude the possibility of the military being able to address other kinds of security challenges, and many of these capabilities are what might be described as fungible forms of military power. This means they have utility beyond conventional war, and a variety of weapons might be used to help save those caught in natural disasters, to support the government in a civil emergency, for example, a pandemic, or a slightly more traditional role within the context of intervention in a civil war. One might argue the extension of war by the military into the political, social and information domains, which new technologies have facilitated in the form of the internet and the smartphone, demonstrates an increasing willingness of the military to look beyond the invention of weapons intended to win battles and this demonstrates an ability to think 'outside the box'. Seen in this light, current Western thinking about future war challenges the existing orthodoxy that dominates debates about the military's relationship with technology, which highlights how it is used to secure their narrow organizational interests.

However, less certain is whether current military thinking goes far enough in terms of addressing the challenges posed by current technological change.

The fourth industrial revolution has spawned a range of technologies that promise transformation across all the sources of social power – ideological, political, economic and military – and this, in turn, will have important consequences for the nature and conduct of war and the cohesion and functioning of the state. The first and most important question is whether the state will be able to retain its monopoly on the use of force within its borders. There is increasing evidence that cheaper but equally potent technologies are emerging which do not require a vast investment to be created and most definitely do not depend on an elaborate but expensive infrastructure to sustain them is 'democratizing' the means of war. Most worrying is the potential that some of these technologies have the capacity to cause casualties on a large scale. In thinking about the probability of someone making the rational decision to take up arms against the state, we need to think first about their motivation to cause harm and then the ability of the state to deter such action. The motivational piece is a complex issue, but the economic consequences of automation and the spread of AI pose a considerable challenge to governments and society. The creation of wealth has played an instrumental role in making political stability and good governance. Also of critical importance is the distribution of this wealth and the opportunities created to promote social mobility.

In contrast, the polarization of wealth creates political division, and leads to the entrenchment of privilege and the manipulation of opportunity so that social mobility ceases to exist in any meaningful sense. These have been persistent problems in even the most open economies today. However, technological innovation threatens to exacerbate and accelerate these trends, which will lead to discontent, which, as has been said, could erupt into different forms of political protest and even violence should the situation become extreme. How does a government with a diminishing tax base support ensure it can satisfy its population's basic needs, and how will it maintain a sufficiently effective state apparatus capable of deterring large-scale violence? This might be addressed through the provision of some form of universal basic income, but the cost of such a programme is hugely expensive. In the case of the United States, the provision of US$12,000 per annum to each US citizen would cost US$3.1 trillion, which equates to 90 per cent of all the money the federal government collected in 2021 (Vesoulis and Abrams, 2021). The big question focuses on how to fund this state subsidy. In the United States and UK, their interpretation of Western economic ideology, with its emphasis on fiscal orthodoxy and commitment to a small state, will make the government reluctant to provide funding on this scale. However, even if it was more generous in its conception of the state's role, it is not clear if the resource could be found to support a comprehensive stimulus package of this size, especially given the other demands likely to face governments in the future. This does not mean the end of the state, but in the same way that it became necessary to reformat its purpose and remit in response to the economic and political challenges it faced in the 1970s, we could face such a moment soon (Bobbitt, 2002, 16–25).

Technology will add fuel to this political and economic bonfire because of how the vast expanse of social media is shaping and reinforcing our views and attitudes. Both state and non-state actors are keenly aware of the power of the internet and how it can be used to shape public opinion by feeding on the resentments and perceived injustices people experience. If we are looking at the world of mass unemployment, it seems likely that much of this resentment will be directed at the state. Governments will also be aware of the potential power of this instrument. They may well seek to imitate Orwell's Oceania in his novel *1984*, but in a world where technology reinforces bias and bounded rationality, the effect of any government propaganda campaign is likely to be resisted, challenged and denied, irrespective of how compelling the evidence is. In sum, technology will exert a profound impact on the conduct of war, but not necessarily in the way Western militaries have assumed. The most important of these is whether the military can rely on a functioning state apparatus and the provision of legitimate government. Technological change is challenging both of these domains, and this is potentially weakening the power of the state. As a result, as we travel inevitably towards the metaverse and all that it has to offer, we could encounter an increasingly fragmented war and state relationship, which causes us to focus less on great power competition and more on addressing the rising threat from within.

Looking further into the future, the internal focus and the tension in the war–state relationship will possibly intensify as non-traditional security challenges begin to exact a toll on the state and exacerbate the possibility of internal and external conflict. In 2021, the US National Intelligence Council produced a detailed forecast of global trends, which also focused on broader non-traditional threats. It noted that climate change, disease, financial crisis and technology disruptions were likely to produce strains on states and societies that could be catastrophic. They also observed that 'the Covid-19 pandemic marks the most significant, singular disruption since World War 2, with health, economic, political, and security implications that will ripple for years to come' (National Intelligence Council, 2021, 2). The IMF estimates that COVID-19 will cost the global economy US\$12.5 trillion by 2024 and exacerbate existing economic and social inequalities within and between states. Most important, a report published in 2022 highlights that global pandemics are likely to be a persistent feature of the future security landscape (Reuters News Agency, 2022). COVID-19 is categorized as a zoonotic disease which has spilt over from wildlife into the human population. It seems about 70 per cent of new viruses emerge from this source. This includes recent scares such as SARS, bird flu, Ebola, HIV and Monkeypox. It is estimated that about three million people die every year from different types of zoonotic disease. Increases in the global population and urbanization mean that human interaction with wildlife is likely to grow, and this increases the chance of new viruses emerging, which might mutate into new pandemics. This explains why novel or exotic viral outbreaks are becoming more frequent (Bernstein et al, 2022). This is clearly one of the great known unknowns of our time, but COVID-19 left behind it a trail of economic devastation. In the case of the United Kingdom, the government was forced to spend over US$580 billion dealing with the effects of the pandemic over the first two years, which was the equivalent of 25 per cent of its GNP, a sum of money one would typically expect to see committed during a time of war. This highlights the multi-faceted nature of the threat facing states. Although it is unlikely that pandemics will cause war, it most definitely leaves a legacy in terms of weakening the state's resilience to deal with follow-on crises. Of these, perhaps the most important is the threat posed by climate change.

In economic terms, the global impact of climate change is predicted to result in a reduction in economic value of 10 per cent. However, in the case of Southeast Asia, the loss will be far more severe, amounting to a 25 per cent reduction in GNP by mid-century if efforts to reduce the increase in greenhouse gases fail to stop temperatures rising above 1.5°C. In the specific case of Malaysia, Thailand and the Philippines, these states could lose between 33 and 36 per cent of their GDP by 2048. If the temperature increases over 2°C then the loss will be even greater, approaching 45 per cent. Even China could see its GDP fall by 15–18 per cent by 2050 if temperatures increase above 2°C.

In contrast, the predicted impact of climate change in North America and Europe is relatively benign. On average it is estimated the United States, Canada and the UK will suffer a 7–8 per cent loss in GDP. Taking into account extreme weather events, the United States is listed as seventh in the league of states least affected by climate change and the UK is fifteenth. In the case of the United States, the principal impact will be increased flooding on the eastern seaboard with increased temperature rises in the Mid-West and South-West United States leading to falls in labour productivity. In the UK, climate change will lead to increased rainfall and storm surges which will cause flash flooding. In both countries, crop yields will be affected, but overall, the impact of these changes will be manageable (Minami Gonzales Wanyu, 2019). More dramatic are the claims made by researchers at the University of East Anglia who estimate that a 35 cm increase in sea levels around the UK over the next 30 years will result in the potential loss of 200,000 homes along vulnerable stretches of the coastline (Warren et al, 2022). If we accept the likely impact of climate change on the GNP of Western states. A fall of 9 per cent of US GDP equates to US$227 billion dollars, which is a third of the current US defence budget. Under these circumstances, fiscal constraints will further limit the capacity of the state to address the challenges it faces and potentially feed into another cycle of violence and instability.

Experts refer to climate change as an accelerator of conflict which operates in conjunction with other factors like poverty and weak governance as a cause of civil wars (DCDC, 2017). However, this view is contested by Wallace-Wells (2019, 124) who observes: 'Wars are not caused by climate change only in the same way that hurricanes are not caused by climate change, which is to say they are made more likely, which is to say the distinction is semantic.'

Recent efforts to measure the relationship between increased temperature and violence revealed that for every half degree of warming, societies will see between a 10 and 20 per cent increase in the likelihood of armed conflict. Extrapolating from this data, Wallace-Wells (2019) claims a four-degree increase in global temperature might produce twice as many wars as happen today. He also refers to another study which revealed that, between 1980 and 2010, 23 per cent of the world's ethnically diverse conflicts began in the months marked by extreme weather disaster. Looking to the future, it is estimated that some 32 countries, all heavily dependent on agriculture, face the risk of rising conflict caused by more extreme weather events produced by climate change. Finally, he also asserts that most wars in history have been over resource conflicts and a combination of climate change and rising population will feed this chain of causation in the future (Wallace-Wells, 2019, 124–7). This depressing picture of the future is confirmed by another study, which predicted a 54 per cent increase in armed conflict and 393,000, more battle deaths in Sub-Saharan Africa as rising temperatures feed civil wars (Burke, 2009). Resource scarcity – for example, diminishing supplies of fresh water – is also linked directly to climate change (Evans, 2011). Access to water is likely to lead to both inter and intra-state conflict. Potential hotspots include the Nile, the Tigris-Euphrates and the Indus, areas that have experienced instability and conflict in the past. Based on the behaviour of violent non-state actors such as Islamic State and al Shabab, it is also possible that both state and non-state actors could weaponize water in areas that are suffering from an acute shortage to achieve their aims.

One way of demonstrating the connection between conflict and climate is to look back rather than forwards in search of much needed evidence to fill this largely speculative vacuum. As the historian Geoffrey Parker observes:

> Most attempts to predict the consequences of climate change extrapolate from recent trends; but another methodology exists. Instead of hitting the fast forward button, we can rewind the tape of history and study the genesis, impact and consequences of past catastrophes. (Parker, 2014, 47)

A paper produced in 2007 by Zhang et al attempted to do precisely this by comparing changing patterns in the Earth's climate and seeing if there is a connection with population increases and decreases and increases and decreases in the occurrence of war. Their initial study focused on China, where they found a strong connection between the occurrence of wars, population change and falling temperatures, which impacted on crop yields, which in turn imposed social and economic stresses on society leading to instability and conflict. These societies were stable before a fall in temperature produced famine. Zhang et al then extended this study to examine climate change across the global and continental geopolitical landmass during the 500-year period known as the Little Ice Age (1400–1900). This expanded study again demonstrated a strong correlation between temperature change, falling agricultural output unrest, war and famine in the northern hemisphere. Of these the most violent was in the 17th century, when most of the northern hemisphere was affected by falling temperatures. The implication here is that food scarcity was a fundamental cause of war:

> A direct cause, in which resource orientated wars erupted as most of the world's population still struggled to satisfy the lower levels of Maslow's Hierarchy of Needs, and an indirect cause, as constrained food resources and economic difficulties stemming from that intensified different social contradictions, that increased the likelihood of war outbreaks. (Zhang et al, 2007, 19218)

The claims made in this study broadly coincide with Parker's (2014) study of climate change and war in the northern hemisphere during the 17th century. His historical survey explores how change in temperature impacted on the political stability of empires and kingdoms located in the northern hemisphere between 1640 and 1690. He demonstrates a strong link between falling temperatures, in this case a fall of 1°C and the crisis that arose across most polities during this period. He makes clear he is not claiming that climate change caused recession, revolution and war. As important here is how states responded to the challenge posed by climate change. As such, climate change did not automatically lead to war and the experience of Japan during this period of the Little Ice Age demonstrates how, even in adversity, while states had no choice but to confront this change, they did have some choice in terms of their response. In contrast to Europe, which was ravaged by one of the most violent and war like periods of its history, best illustrated by the Thirty Years War, Japan experienced rapid demographic, agricultural and urban growth and no wars (Parker, 2014, 876).

The historical evidence shows that climate change has played both an instrumental and a facilitating role in the promotion of conflict in the past. The principal difference today is that first climate change is clearly linked to human activity, which implies there is a human solution available that focuses on creating a zero-carbon economy. Second, temperatures are going up, not down, which means there could be climate change winners as well as losers – for example, the opportunities created by the melting of the ice caps in the north will create clear economic opportunities for Canada and Russia. Finally, the magnitude of the change will be global in scale. If climate change is not addressed, it's likely to overlap in terms of time with the ongoing effects caused by the fourth industrial revolution, and this too will help shape the causes and conduct of war, and the role of the state. These elements when combined suggest the 21st century will not be the bright shiny future imagined by technological determinists but something darker and more dystopian.

Within this more volatile setting technology is creating a new policy logic and a new grammar of war which is empowering new political forms and new means to wage war. The combined effect of these changes could weaken the historic link forged between technology, war and the state, and damage the power of the state in terms of its ability to generate the power to create the resilience needed to address future large scale crises that are looming on the horizon.

References

Allison, G. (2018) *Destined for War: Can America and China Escape the Thucydides' Trap*. London: Scribe.

Angstrom, J. and Widen, J.J. (2015) *Contemporary Military Theory*. Abingdon: Routledge.

Arreguin Toft, I. (2001) 'How the Weak Win Wars: A Theory of Asymmetric Conflict', *International Security*, 26(1): 93–128.

Augustine, N. (1997) *Augustine's Laws*. Reston, VA: American Institute of Aeronautics.

Azhar, A. (2021) *Exponential*. London: Random House Business.

Banerjee, A. (2021) 'Arms and the Man: Strategic Trade Control Challenges of 3D Printing', *International Journal of Nuclear Security*, 4(1).

Baxter, M. and de Salaberry, J. (2020) *Living in the Age of the Jerk*. Ottawa: Techopia.

Bean, R. (1973) 'War and the Birth of the Nation State', *Journal of Economic History*, 33(1): 203–21.

Beissinger, M.R. (2022) *The Revolutionary City: Urbanization and the Global Transformation of Rebellion*. Princeton, NJ: Princeton University Press.

Berghan, V.R. (1984) *Militarism*. Cambridge: Cambridge University Press.

Bernstein, A.S., Ando, A.W., Loch-Temzelides, T., Vale, M.M., Li, B.V., Li, H. et al (2022) 'The Costs and Benefits of Primary Prevention of Zoonotic Pandemics', *Sciences Advances*, 8(5).

Betz, D. (2012) 'Cyberpower in Strategic Affairs', *Journal of Strategic Studies*, 35(5): 695.

Biddle, S. (2022) 'Ukraine and the Future of Offensive Maneuver', *War on the Rocks*, November.

Bittleston, M. (1990) Cooperation or Competition? Defence Procurement Options for the 1990s. Adelphi Paper Series 250.

Bitzinger, R. (2016) 'Is China's Space Program Rocketing Past America?' *National Interest*, 10 May.

Black, J. (1991) *A Military Revolution? Military Change in European Society 1550–1800*. London: Macmillan.

Black, J. (1994) *European Warfare, 1660–1815*. London: UCL Press.

Black, J. (2004) *Rethinking Military History*. Abington: Routledge.

Black, J. (2013) *War and Technology*. Bloomington, IN: Indiana University Press.

Bobbitt, P. (2002) *The Shield of Achilles: War, Peace, and the Course of History*. New York: Knopf.

Bolia, R.S. (2004) 'Overreliance on Technology in Warfare: The Yom Kippur War as a Case Study', *Parameters*, 34(2).

Bond, B. (1977) *Liddell Hart. A Stud of His Military Thought*. London: Cassell.

Brodie, B. (2014) 'The Absolute Weapon', in T.G. Mankhen, and J.A. Maiolo (eds) *Strategic Studies: A Reader*. Second edition. London: Routledge.

Brodie, B. and Brodie, F.M. (1973) *From Crossbow to H Bomb: The Evolution of the Weapons and Tactics of Warfare*. Bloomington, IN: Indiana University Press.

Brown, T. (2018) 'Space Land Battle', *Military Review*, November/December.

Burke, M. (2009) 'Warming Increases the Risk of Civil War in Africa', *Proceedings of the National Academy of Sciences*, 106(4).

Buzan, B. and Waever, O. (2003) *Regions and Powers: The Structure of International Security*. Cambridge: Cambridge University Press.

Buzan, T. (1987) *Military Technology and International Relations*. London: Macmillan.

Byrne, J., Somerville, G., Byrne, J., Watling, J., Reynolds, N. and Baker, J. (2022) *Silicon Lifeline: Western Electronics at the Heart of Russia's War Machine*. London: RUSI, August.

Cadwaller, C. (2022) 'Social Media Turn on Putin, the Past Master', *Daily Telegraph*, 6 March.

Caverly, J.D. (2009/10) 'The Myth of Military Myopia: Democracy, Small Wars, and Vietnam', *International Security*, 34(3): 119–57.

Chalmers, M. (1980) *Paying for Defence Military Spending and British Decline*. London: Pluto Books.

Chin, W. (2019) 'Technology, War and the State Past, Present and Future', *International Affairs*, 94(4).

Chin, W.A. (2004) *British Weapons Acquisition Policy and the Futility of Reform*. Aldershot: Ashgate.

Chueng, T.M. (2022) *Innovate to Dominate: The Rise of the Chinese Techno-Security State*. Ithaca, NY: Cornell University Press.

Cimbala, S. (2012) *Clausewitz and Escalation Classical Perspectives on Nuclear Strategy*. Abingdon: Routledge.

Cohen, E. (2004) 'Change and Transformation in Military Affairs', *Journal of Strategic Studies*, 27(3).

Cohen, E.A. (1996) 'A Revolution in Warfare', *Foreign Affairs*, 75(2): 37, 10.2307/20047487.

Coker, C. (2014) *Future War*. Malden, MA: Polity Press.

Congressional Research Service (2020) *Russian Armed Forces: Capabilities*. 30 June.

Cornish, P. and Donaldson, K. (2017) *2020: World of War*. London: Hodder & Stoughton.

Covini, M.N. (2001) 'The Political and Military Bonds in the Italian State System, Thirteenth to Sixteenth Centuries' in P. Contamine (ed) *War and Competition Between States*. London: Clarendon Press.

Cox, D. (2021) 'Artificial Intelligence and Multi-Domain Operations: A Whole Nation Approach Key to Success', *Military Review*, May/June.

Cramer, C. (2006) *Civil War is Not a Stupid Thing: Accounting for Violence in Developing Countries*. London: Hurst & Company.

CSIS (2023) 'What does China Really Spend on its Military, www. csis.org/analysis/what-does-china-really-spend-its-military#:~:text= The%20Chinese%20government%20announces%20defense,1.45%20 trillion%20(%24229.6%20billion)

Darby, C. and Sewell, S. (2021) 'The Innovation Wars America's Eroding Technological Advantage', *Foreign Affairs*, March/April.

DCDC (2017) *Global Strategic Trends Programme, 2007–2036*, www.gov.uk/ government/publications/global-strategic-trends-the-future-starts-today

DCDC (2018) *Global Strategic Trends Programme, 2040*, chrome-extension:// efaidnbmnnnibpcajpcglclefindmkaj/https://assets.publishing.service.gov. uk/government/uploads/system/uploads/attachment_data/file/1075981/ GST_the_future_starts_today.pdf

DCDC (2021) *The Orchestration of Military Effects*, https://assets.publishing. service.gov.uk/government/uploads/system/uploads/attachment_data/ file/970529/20210316-OMSE_new_web-O.pdf

Delbruck, H. (1990) *History of the Art of War: The Dawn of Modern Warfare, Volume 4*. Lincoln, NE: Nebraska University Press.

Diamandis, P.H. and Kotler, S. (2020) *The Future Is Faster than You Think: How Converging Technologies Are Transforming Business, Industries, and Our Lives*. New York: Simon & Schuster.

Dobbs, R., Manyika, J. and Woetzel, J. (2015) *No Ordinary Disruption: The Four Global Forces Breaking All the Trends*. New York: Public Affairs.

Doughty, R.A. (2014) *The Breaking Point: Sedan and the Fall of France, 1940*. Mechanicsburg, PN: Stackpole Books.

Downing, B. (1992) *The Military Revolution and Political Change Origins of Democracy and Autocracy in Early Modern Europe*. Princeton, NJ: Princeton University Press.

Drum, K. (2018) 'Tech World Welcome to the Digital Revolution', *Foreign Affairs*, 97(4).

Duffy, C. (1979) *Siege Warfare: The Fortress in the Early Modern World 1494–1660*. London: Routledge.

Echevarria, A. (2022) 'Putin's Invasion of Ukraine in 2022: Implications for Strategic Studies', *Parameters*, 52(2).

Economist, The (2020) 'Space Weapons: The High Ground', 1 February.

Economist, The (2021) 'Nominal Spending Figures Understate China's Military Might', 1 May.

Ede, A. (2019) *Technology and Society A World History*. Cambridge: Cambridge University Press.

Elliot, L. (2015) 'Robots Threaten 15m Jobs, Says Bank of England Chief Economist', *The Guardian*, 12 November.

Ellis, J. (1975) *The Social History of the Machine Gun*. London: Croom Helm.

Ellman, J., Samp, L. and Coll, G. (2017) *Assessing the Third Offset*. Washington DC: CSIS.

Elstub, J. (1969) *On The Productivity of the National Aircraft Effort*. London: HMSO.

Erwin, S. (2018) 'US Intelligence: Russia and China Will Have Operational Anti Satellite Weapons in a Few Years', *Space News*, 14 February.

European Parliamentary Research Service (2022) *Russia's War on Ukraine: Timeline of Cyber-Attacks*. Brussels: European Union.

Evans, A. (2011) *Resource Scarcity, Climate Change and the Risk of Violent Conflict*. World Development Bank Report, Background paper, https://openknowledge.worldbank.org/handle/10986/9191

Farrell, T. (2005) *The Norms of War: Cultural Beliefs and Modern Conflict*. Boulder, CO: Lynne Rienner Publishers.

Fazal, T.M. (2018) 'The Return of Conquest? Why the Future Global Order Hinges on Ukraine', *Foreign Affairs*, 101(3): 20–7.

Ferguson, J., Traynor, L. and Yallop, H. (2017) *Arms and Armour of the First World War*. Leeds: Royal Armouries Museum.

Fiztgerald, B and Sayler, K. (2022) 'Emerging Technologies: Background and Issues for Congress', Congressional Research Service R46458, 1 November.

Fleming, C. (2009) 'Old or New Wars? Debating a Clausewitzian Future', *Journal of Strategic Studies*, 32(2).

Ford, M. (2022) *Rule of the Robots: How Artificial Intelligence Will Transform Everything*. New York: Basic Books.

Freedman, L. (2003) *The Evolution of Nuclear Strategy*. Third edition. Basingstoke: Palgrave.

Fridman, O. (2018) *Russian Hybrid Warfare: Resurgence and Politicization*. Oxford: Oxford University Press.

Friedberg, A.L. (2000) 'The United States and the Cold War Arms Race', in O.A. Westad (ed) *Reviewing the Cold War Approaches, Interpretations, Theory*. London: Routledge.

Fukuyama, F. (2011) *The Origins of Political Order: From Prehuman Times to the French Revolution*. London: Profile Books.

Fukuyama, F. (2015) *Political Order and Political Decay from the Industrial Revolution to the Globalization of Democracy.* London: Profile Books.

Fuller, J.F.C. (1998) *Influence of Armament on History: From the Dawn of Classical Warfare to the End of the Second World War.* Boston, MA: Da Capo Press.

Galbraith, J.K. (1967) *The New Industrial State.* Princeton, NJ: Princeton University Press.

Galeotti, M. (2022) *The Weaponisation of Everything: A Field Guide to the New Way of War.* New Haven, CN: Yale University Press.

Gallaghar, G.W. (1977) *The Confederate War.* Cambridge, MA: Harvard University Press.

Garrett, L. (2013) 'Biology's Brave New World: The Promise and Perils of the Syn Bio Revolution', *Foreign Affairs,* 92(6).

Gerth, H.H. and Mills, C.W. (1946) *From Max Weber: Essays in Sociology.* New York: Oxford University Press.

Gibson, J.W. (1986) *The Perfect War: Technowar in Vietnam.* New York: Atlantic Monthly Press.

Glantz, D.M. (2012) *Soviet Military Deception in the Second World War.* Hoboken, NY: Taylor and Francis.

Glantz, D.M. (2015) *When Titans Clashed How The Red Army Stopped Hitler.* Lawrence, KS: Kansas University Press.

Glanz, J. and Hernandez, M. (2022) How Ukraine Blew up a Key Russian Bridge, *New York Times,* 17 November, www.nytimes.com/interactive/2022/11/17/world/europe/crimea-bridge-collapse.html

Goldfarb, A. and Lindsay, J.R. (2022) 'AI and the Human Context of War', *National Interest,* 30 April. https://nationalinterest.org/blog/tech land-when-great-power-competition-meets-digital-world/artificial-intelligence-and-human

Gorgol, J.F. (1972) *Military Industrial Firm: A Practical Theory and Model.* New York: Praeger.

Graham, S. (2011) *Cities Under Siege: The New Military Urbanism.* London: Verso.

Gray, C. (2005) *Another Bloody Century.* London: Weidenfeld and Nicholson.

Guderian, H. (2001) *Achtung-Panzer! The Development of Armoured Forces, Their Tactics and Operational Potential.* London: Brockhampton.

Gurr, T.R. (2015) *Political Rebellion: Causes, Outcomes and Alternatives.* Abingdon: Routledge.

Hables Gray, C. (2013) *Postmodern War: The New Politics of Conflict.* London: Routledge.

Hacker, B.C. (1994) 'Military Institutions, Weapons, and Social Change: Toward a New History of Military Technology', *Technology and Culture,* 35(4): 768, 10.2307/3106506.

Hanson, V.D. (2001) *Carnage and Culture Landmark Battles in the Rise of Western Power.* New York: Doubleday.

Harari, N.Y. (2017) *Homo Deus: A Brief History of Tomorrow.* London: Vintage.

Hartcup, G. (1993) *The Silent Revolution: The Development of Conventional Weapons 1945–85.* London: Brassey's.

Hastings, M. (2009) *Finest Years: Churchill as Warlord 1940–45.* London: Harper Press.

Haughey, C. (2021) 'Cyber Warfare: What to Expect in 2022', *Security Intelligence*, 22 December.

Heilbroner, R.L. (1967) 'Do Machines Make History?' *Technology and Culture*, 8(3): 335, 10.2307/3101719.

Hicks, M. and Devaraj, S. (2015) Myth and Reality in Manufacturing in America, www.nist.gov/system/files/documents/mep/data/Mfg Reality-1.pdf

Higgs, R. (ed) (1990) *Arms, Politics, and the Economy Historical and Contemporary Perspectives.* New York: Holmes and Meier.

Hird, K., Stepanenko, K. and Clark, M. (2022) Russian Offensive Campaign Assessment, June 9, www.understandingwar.org/backgrounder/russian-offensive-campaign-assessment-june-9

Hirst, P. (2001) *War and Power in the 21st Century.* Cambridge: Polity Press.

HM Government (2021) *Global Britain in a Competitive Age: The Integrated Review of Security, Defence, Development and Foreign Policy.* CP 403, London: HMSO, March.

Holden-Reid, B. (1987) *JFC Fuller Military Thinker.* Basingstoke: Palgrave Macmillan.

Holsti, K. (1991) *Peace and War: Armed Conflicts and International Order 1648–1989.* Cambridge: Cambridge University Press.

Howard, M. (2009) *War in European History.* Oxford: Oxford University Press.

Huntington, S.P. (1985) *The Soldier and the State*, Cambridge, MA: Harvard University Press.

Huntington, S.P. (2005) 'Arms Races: Prerequisites and Results', in R.K. Betts (ed) *Conflict After the War on Terror.* London: Pearson Longman.

Hutton, W. (2007) *The Writing on the Wall China and the West in the 21st Century.* London: Abacus.

Iansiti, M. and Lakhani, K. (2020) *Competing in an Age of AI Strategy Leadership When Algorithms and Networks Run the World.* Boston, MA: Harvard Business School.

Ignatieff, M. (2001) *Virtual War: Kosovo and Beyond*, London: Picador.

Johnson, R. (2022) 'Dysfunctional Warfare: The Russian Invasion of Ukraine', *Parameters*, 52(2).

Kagan, R. (2008) *The Return of History and the End of Dreams.* New York: Alfred Knopf.

Kai Fu Lee (2019) *AI Superpowers, China, Silicon Valley, and the New World Order.* New York: Mariner Books.

Kaldor, M. (1998) *New and Old Wars: Organized Violence in a Global Era.* Cambridge: Polity Press.

Kaplan, R.D. (2012) *The Revenge of Geography: What the Map Tells Us about Coming Conflicts and the Battle against Fate.* New York: Random House.

Keegan, J. (2014) *The Face of Battle: A Study of Agincourt, Waterloo and the Somme,* London: The Bodley Head.

Kemp, R. (2022) 'Putin has Regained the Military Initiative', *Daily Telegraph,* 30 May.

Kershaw, I. (2000) *Hitler: Nemesis 1936–45 (Book 2).* London: Penguin.

Kier, E. (2017) *Imagining War French and British Military Doctrine Between the Wars.* Princeton, NJ: Princeton University Press.

Kilcullen, D. (2020) *The Dragons and the Snakes How the Rest Learned to Fight the West.* London: Hurst & Company.

Kirkpatrick, D.L.I. and Pugh, P.G. (1985) 'Towards the Starship Enterprise – are the Current Trends in Defence Unit Costs Inexorable?', *Journal of Cost Analysis,* 2(1): 59–80.

Kissinger, H. (2018) 'How the Enlightenment Ends', *The Atlantic,* 1 June, www.theatlantic.com/magazine/archive/2018/06/henry-kissinger-ai-could-mean-the-end-of-human-history/559124/

Knox, M. and Murray, W. (eds) (2001) *The Dynamics of a Military Revolution 1300–2050.* Cambridge: Cambridge University Press.

Kofman, M. and Connolly, R. 'Why Russian Military Expenditure is Much Higher Commonly Understood (As is China's)', *War on the Rocks,* 16 December 2019.

Krepinevich, A. (1994) 'From Cavalry to the Computer, The Pattern of Military Revolutions', *The National Interest,* 37(4): 30–42.

Kurth Cronin, A. (2022) 'Biology's Tectonic Shifts and Novel Risks', *CTC Sentinel,* 15(5), https://ctc.westpoint.edu/biologys-tectonic-shifts-and-novel-risks/

Kurzweil, R. (2014) 'The Law of Accelerating Returns', www.kurzweilai.net/the-law-of-accelerating-returns

Laswell, H. (1997) *Essays on the Garrison State,* edited by J. Stanley. New Brunswick, NJ: Transaction Press.

Latiff, R.H. (2018) *Future War: Preparing for the New Global Battlefield.* New York: Vintage Books.

Lee, K.-F. (2018) *AI Superpowers: China, Silicon Valley and the New World Order.* New York: Houghton Mifflin Harcourt.

Lee, R. (2022) 'The Tank is Not Obsolete, and Other Observations about the Future of Combat', *War on the Rocks,* 6 September.

Lendon, B. (2022) 'Putin Can Call Up all the Troops he Wants, but Russia Can't Train Them', *CNN World* News, 22 September.

Levy, Y. (2012) 'A Revised Model of Civilian Control of the Military: The Interaction between the Republican Exchange and the Control Exchange', *Armed Forces and Society*, 38(4): 529–56.

Lind, W.S. (1989) 'The Changing Face of War: Into the Fourth Generation', *Marine Corps Gazette*, October: 22–26.

Lind, W.S. and Thiele G.A. (2015) *4th Generation Warfare Handbook*. Kouvola: Castalia House.

Luzin, P. (2022) 'Russia's Skyrocketing Defense Spending, 2022-23', Jamestown Foundation, *Eurasia Daily Monitor*.

Lyall, J. and Wilson, I. (2009) 'Rage against the Machines: Explaining Outcomes in Counterinsurgency Wars', *International Organization*, 63(1).

Mack, A. (1975) 'Why Big Nations Lose Small Wars: The Politics of Asymmetric Conflict', *World Politics*, 27(2): 175–200.

Macmillan, M. (2018) 'War and Humanity', The Reith Memorial Lectures, *BBC Radio 4*, 26 June.

Macmillan, M. (2020) *War: How Conflict Shaped The World*. London: Profile Books.

Mann, M. (1983) *The Sources of Social Power. Volume 1: A History of Power from the Beginning to AD 1760*. New York: Cambridge University Press.

Mann, M. (1986) *The Sources of Social Power. Volume 1: A History of Power from the Beginning to AD 1760*. Cambridge: Cambridge University Press.

Mann, M. (2013) *The Sources of Social Power. Volume 4: Globalizations, 1945– 2011*. Cambridge: Cambridge University Press.

Manyika, J., Madgavakr, A., Woetzel J., Smit, S. and Abudaal, A. (2020) *The Social Contract in the 21st Century Outcome so Far for Workers, Consumers, and Savers in Advanced Economies*. New York: McKinsey Global Institute.

Marr, B. (2022) *Business Trends in Practice the 25+ Trends that are Redefining Organizations*. Oxford: Wiley Blackwell.

Mathis, J.N. (2008) 'USJFCOM Commander's Guidance for Effects Based Operations', *Parameters,* Autumn 2008.

McAfee, A. and Brynjolfson, E. (2014) *The Second Machine Age: Work, Progress in Times of Brilliant Technologies*. New York: Norton.

McInnes, C. (2002) *Spectator Sport War: The West and Contemporary Conflict* Boulder, CO: Lynne Reiner.

McMaster, H.R. (2016) 'Eagle Troop and the Battle of 73 Easting', *Strategy Bridge*, thestrategybridge.org

McNeill, W.H. (1982) *The Pursuit of Power: Technology, Armed Force, and Society since A.D. 1000*. Chicago, IL: University of Chicago Press.

Mearsheimer, J. (2017) *Conventional Deterrence*. Ithaca, IL: Cornell University Press.

Merom, G. (2003) *How Democracies Lose Small Wars: State, Society, and the Failures of France in Algeria, Israel in Lebanon, and the United States in Vietnam*. Cambridge: Cambridge University Press.

Misa, T.J. (1988) 'How Machines Make History, and How Historians (and Others) Help Them to Do So', *Science, Technology, & Human Values*, 13(3-4): 308–31.

Morillo, S. (1999) 'The Age of Cavalry Revisited', in D.J. Kagay and L.J.A. Villalon (eds) *The Circle of War in the Middle Ages*. Woodbridge: Boydell & Brewer.

Mowery, D. and Rosenberg, N. (1989) *Technology and the Pursuit of Economic Growth*. Cambridge: Cambridge University Press.

Mueller, J. (1989) *Retreat from Doomsday: The Obsolescence of Major War*. New York: Basic Books.

Murray, W. and Millett, A.R. (eds) (1996) *Military Innovation in the Interwar Period*. Cambridge: Cambridge University Press.

Nagl, J.A. (2005) *Learning to Eat Soup with a Knife: Counterinsurgency Lessons from Malaya and Vietnam*. Chicago, IL: Chicago University Press.

Naím, M. (2014) *The End of Power: From Boardrooms to Battlefields and Churches to States, Why Being in Charge Isn't What It Used to Be*. New York: Basic Books.

National Intelligence Council (2021) *Global Trends 2040*. Washington DC: NIC.

Naveh, S. (1997) *The Pursuit of Military Excellence The Evolution of Operational Theory*. Abingdon: Routledge.

Noble, R.K. (2013) 'Keeping Science in the Right Hands', *Foreign Affairs*, 92(6).

Nurkin, T. and Rodriguez, S. (2019) *A Candle in the Dark: US National Security Strategy for Artificial Intelligence*. Washington, DC: Atlantic Council of the United States.

O'Hanlon, M. (2018) 'The Role of AI in Future Warfare', *Brookings Institution*, 29 November, brookings.edu/research/ai-and-future-warfare/

Oliver, M. (2022) 'How British Tech is Driving Putin's War Machine', *Daily Telegraph*, 8 May.

Organski, A.F.K. (1968) *World Politics*. Ann Arbor, MI: Michigan University Press.

Osgood, R.E. (1957) *Limited War: The Challenge to American Strategy 5th Edition*. Chicago, IL: Chicago University Press.

Overy, R. (1995) *Why the Allies Won*. London: Jonathan Cape.

Page, L. (2022) 'Why the Invasion of Ukraine Spells of Modern Tank Warfare, *The Telegraph*, 29 May, www.telegraph.co.uk/business/2022/05/29/modern-warfare-myth-tank-supremacy-has-blown-ukraine/

Parker, G. (2014) *Global Crisis: War, Climate Change and Catastrophe in the Seventeenth Century*. New Haven, CN: Yale University Press.

Parker, G. (2016) *The Military Revolution: Military Innovation and the Rise of the West, 1500–1800*. Cambridge: Cambridge University Press.

Parrott, D. (2012) *The Business of War Military Enterprise and Military Revolution in Early Modern Europe*. Cambridge: Cambridge University Press.

Pavel, B. and Vekatram, K. (2021) Facing the Future of Bio Terrorism, *The Atlantic Council*, 7 September, www.atlanticcouncil.org/commentary/article/facing-the-future-of-bioterrorism/

Pearton, M. (1982) *The Knowledgeable State: Diplomacy, War and Technology since 1830*. London: Burnett Books.

Peck, M.J. and Scherer, F.M. (1962) *The Weapons Acquisition Process An Economic Analysis*. Boston, MA: Harvard University Press.

Pinkstone, J. (2022) 'Why Vladimir Putin's Propaganda "Nonsense" Failed to Catch on in Ukraine', *Daily Telegraph*, 3 May.

Posen, B. (1984) *The Sources of Military Doctrine: France, Britain and Germany Between the World Wars*. Ithaca, NY: Cornell University Press.

Price, A. (1967) *Instruments of Darkness*. London: William Kimber.

Price, R. and Tannenworld N. (1996) 'Norms and Deterrence: The Nuclear and Chemical Weapons Taboo,' in P. Katzenstein (ed) *The Culture of National Security: Norms and Identity in World Politics*. New York: Columbia University Press.

Prokopenko, A. (2022) 'Could Russia Move First to Halt Oil Exports to Europe?', *Carnegie Endowment for International Peace*, 30 May, https://carnegieendowment.org/politika/87209

Pugh, P.G. (1993) 'The Procurement Nexus', *Defense Economics*, 4(2):179-94.

PWC (2018) Will Robots Really Steal Our Jobs? *How Will Automation Impact on Jobs*. www.pwc.co.uk/economic-services/assets/international-impact-of-automation-feb-2018.pdf

Qiao, L. and Wang X. (2015) *Unrestricted Warfare*. Brattleboro, VT: Vt Echo Point Books & Media.

Quigley, C. (2013) *Weapons Systems and Political Stability: A History*. New York: Dauphin Publications.

Reuters News Agency (2022) 'MF sees cost of COVID pandemic rising beyond $12.5 trillion estimate', www.reuters.com/business/imf-sees-cost-covid-pandemic-rising-beyond-125-trillion-estimate-2022-01-20/

Ribot-Garcia, L. (2001) 'Types of Armies: Early Modern Spain' in P. Contamine (ed) *War and Competition Between States*. London: Clarendon Press.

Rid, T. (2021) *Active Measures: The Secret History of Disinformation and Political Warfare*. New York: Picador.

Roberts, M. (1995) 'The Military Revolution, 1560–1660', in C. Rogers (ed) *The Military Revolution Debate: Readings on the Military Transformation of Early Modern Europe*. Abingdon: Taylor & Francis, pp 12–35.

Robinson, L., Helmus, T.C., Cohen, R.S., Nader, A., Rodin, A., Magnuson, M. et al (2018) *Modern Political Warfare Current Practices and Possible Responses*. Santa Monica: RAND.

Rogers, C.J. (1993) 'The Military Revolutions of the Hundred Years' Wars', *Journal of Military History*, 57(2): 241–78.

Russ, D. (2015) 'The Robots are Coming', *Foreign Affairs*, 94(4): 2–6.

Sabbagh, D. (2022) 'Five Predictions for the Next Six Months of the War in Ukraine', *The Guardian*, 24 August.

Samuel, J. (2022) 'Robots are not Destroying all our Jobs. They are the Solution to Our Economic Misery', *Daily Telegraph*, 18 June.

Scales, R. (2019) 'Tactical Art in Future Wars', War on the Rocks, warontherocks.com/2019/03/tactical-art-in-future-wars/

Scharre, P. (2018) *Army of None: Autonomous Weapons and the Future of War*. New York: W.W. Norton & Company.

Schatzberg, E. (2018) *Technology: A Critical History of a Concept*. Chicago, IL: Chicago University Press.

Schulke, M. (2022) *Technological, Organizational, and Strategic Change beyond Conventional War*. Ann Arbor, MI: University Michigan Press.

Schumpeter, J.A. (1994) *Capitalism, Socialism and Democracy*. London: Routledge.

Schwab, K. (2016) *The Fourth Industrial Revolution*. New York: Crown Business.

Shaw, M. (2005) *The New Western Way of War Risk Transfer War and its Crisis in Iraq*. Cambridge: Polity Press.

Shimko, K. (2010) *The Iraq Wars and America's Military Revolution*. Cambridge: Cambridge University Press.

Singer, P.W. (2009) *Wired For War: The Robotics Revolution and Conflict in the 21st Century*. New York: Penguin Press.

Singer, P.W. and Brooking, E.T. (2018) *Like War The Weaponization of Social Media*. New York: Houghton and Mifflin.

Smith, D. (1980) *The Defence of the Realm in the 1980s*. London: Croom Helm.

Smith, R. (2005) *War Amongst the People*. London: Penguin Books

Spruyt, H. (1994) *The Sovereign State and its Competitors*. Princeton, NJ: Princeton University Press.

Statista (2023) Number of Ukrainian Refugees, www.statista.com/statistics/1312584/ukrainian-refugees-by-country/#:~:text=Nearly%202.9%20million%20refugees%20from,refugees%20were%20registered%20across%20Europe

Stephanko, K., Kagan, W.F., Barros, G., Clark, M., Mappes, G. (2022) Institute for the Study of War, Russian Offensive Campaign Assessment, 28 June, www.understandingwar.org/backgrounder/russian-offensive-campaign-assessment-june-28

Stern, J. and Berger, J.M. (2015) 'Isis and the Foreign Fighter Phenomenon', *The Atlantic*, 8 March.

Stiglitz, J.E. and Korinek, A. (2020) *Artificial Intelligence, Globalization, And Strategies For Economic Development*. Working Paper 28453, National Bureau of Economic Research, www.nber.org/papers/w28453

Stockholm International Peace Research Institute (2022) 'World Military Expenditure passes $2 Trillion for the First Time', www.sipri.org/media/press-release/2022/world-military-expenditure-passes-2-trillion-first-time

Strachan, H. (1983) *European Armies and the Conduct of War*. London: Routledge.

Stritmatter, K. (2019) *We Have Been Harmonised: Life in China's Surveillance State*. London: Old Street Publishing.

Summers, H. (2007) *American Strategy in Vietnam: A Critical Analysis*. New York: Dover Publications.

Susskind, J. (2018) *Future Politics: Living Together in a World Transformed by Tech*. Oxford: Oxford University Press.

Tarar, A. (2016) 'A Strategic Logic of the Military Fait Accomplis', *International Studies Quarterly*, 60:4.

Terraine, J. (1988) *The Right of the Line: The Royal Air Force in the European War 1939–45*. Ware: Wordsworth Editions.

Tetlock, P.E. and Gardner, D. (2015) *Superforecasting the Art and Science of Prediction*. London: Random House.

Thomas, J. (2022) 'West Hits Vladimir Putin's Fake News Factories with Wave of Sanctions', *The Observer*, 20 March.

Tilly, C. (1985) 'War Making and State Making as Organized Crime', in P. Evans, D. Rueschemeyer and T. Skocpol (eds) *Bringing the State Back In*. Cambridge: Cambridge University Press, pp 169–87.

Tilly, C. (1992) *Coercion, Captial and European States, AD 990–1992*. Oxford: Blackwells.

Tin-Bor Hui, V. (2005) *War and State Formation in Ancient China and Early Modern Europe*. New York: Cambridge University Press.

Todd, D. (1988) *Defence Industries: A Global Perspective*. London: Routledge.

Toffler, A. and Toffler, H. (1993) *War and Anti War Survival at the Dawn of the 21st Century*. New York: Little Brown and Company.

Triandafilov, V.K. (1994) *The Nature of the Operations of Modern Armies*. Abingdon: Routledge.

Tynan, D. (2018) '"I Wouldn't Waste my Time": Firearms Experts Dismiss Flimsy 3D-Printed Guns', *The Guardian*, 1 August.

UK Ministry of Defence (2021) *Global Strategic Trends – The Future Starts Today*, DCDC.

UK Ministry of Defence (2022) *UK Defence Doctrine*, Joint Doctrine 0–01, London: Ministry of Defence.

Urban, M. (2015) *The Edge: Is Military Dominance of the West Coming to an End?* London: Little Brown.

US Army (2018) *Multi Domain Operations 2028*. TRADOC Pamphlet 525–3-1.

Vagts, A. (1937) *A History of Militarism: Romance and Realities of a Profession*. New York: WW Norton and Company.

Vagts, A. (1957) *A History of Militarism*. New York: Meridian Books.

van Creveld, M. (1989) *Technology and War from 2000 B.C. to the Present*. New York: Free Press.

van Creveld, M. (1991) *On Future War*. London: Brassey's.

van Creveld, M. (2009) *The Rise and Decline of the State*. Cambridge: Cambridge University Press.

van Creveld, M. (2016) *Pussycats: Why the Rest Keeps Beating the West – and What Can Be Done about It*. Mevasseret Zion, Israel: DLVC Enterprises.

van Creveld, M. (2020a) *Command in War*. Cambridge, MA: Harvard University Press.

van Creveld, M. (2020b) *Seeing into the Future: A Short History of Prediction*. London: Reaktion Books.

Vesoulis, A. and Abrams, A. (2021) 'Inside the Nations Largest Guaranteed Income Experiment,' *Time Magazine*, 16 September, https://time.com/6097523/compton-universal-basic-income/

Vlankancic, P.J. (1992) *Marshal Tukhachchevsky and the Deep Battle: An Analysis of Operational Level Soviet Tank and Mechanized Doctrine 1935–45*. AUSA, Institute of Land Warfare No. 14, November.

von Bernhardi, F. (2015) *Germany and the Next War*. Palala Press.

von Clausewitz, C. (1976) *On War*, edited by M. Howard and P. Paret. Princeton, NJ: Princeton University Press.

Vredeling, H. (1987) *Towards A Stronger Europe*. Brussels: IEPG.

Wall Street Journal (2018) 'The Future of Everything: How AI is Augmenting Therapy', podcast, www.wsj.com/podcasts/wsj-the-future-of-everything/how-ai-is-augmenting-therapy/810a7099-0cc3-4e03-8148-dd87c3673152

Wallace, B. (2022) Secretary of State for Defence: statement to the House of Commons on the war in Ukraine, 25 April. https://www.gov.uk/government/speeches/defence-secretary-statement-to-the-house-of-commons-on-ukraine-25-april-2022

Wallace-Wells, D. (2019) *The Uninhabitable Earth: Life After Warming*. London: Penguin.

Walton, S.A. (2019) 'Technological Determinism(s) and the Study of War', *Vulcan*, 7(1)5: 4–18, 10.1163/22134603-00701003.

Warren, R., Andrews, O., Brown, S., Colón-González, F.J., Forstenhäusler, N., Gernaat, D.E.H.J. et al (2022) 'Quantifying Risks Avoided by Limiting Global Warming to 1.5 or 2°C above Pre-industrial Levels', *Climatic Change*, 172(39).

Watling, J. and Reynolds, N. (2022) *Ukraine at War Paving the Road from Survival to Victory*. RUSI Special Report, July, https://rusi.org/explore-our-research/publications/special-resources/ukraine-war-paving-road-survival-victory

Weinberg, G.L. (1995) *A World at Arms: A Global History of World War II*. New York: Cambridge University Press.

Wetzel, T. (2022) Ukraine air war examined. A glimpse into the future of air warfare, *Atlantic Council*, 30 August, www.atlanticcouncil.org/content-series/airpower-after-ukraine/ukraine-air-war-examined-a-glimpse-at-the-future-of-air-warfare

White, L. (1962) *Medieval Technology and Social Change*. London: Oxford University Press.

Winner, L. (1977) *Autonomous Technology: Technics-out-of-control as a Theme in Political Thought*. Cambridge, MA: MIT Press.

Wolf, M. (2015) 'Same As It Ever Was', *Foreign Affairs*, 94(4): 18.

World Economic Forum (2020) *The Future of Jobs*, https://www.weforum.org/reports/the-future-of-jobs-report-2020

Zabrodskyi, M., Watling, J., Danylin, O.V. and Reynolds, N. (2022) 'Preliminary Lessons in Conventional Warfighting from Russia's Invasion of Ukraine: February-July 2022', *RUSI*, 30 November.

Zhang, D., Brecke, P., Lee, H.F., and Zhang, J. (2007) 'Global Climate Change, War, and Population Decline in Recent Human History', *PNAS*, 104(49).

Zimmerman, D. (2019) 'Neither Catapults nor Atomic Bombs', *Vulcan*, 7(1): 45–61, 10.1163/22134603-00701005.

Zuboff, S. (2019) *The Age of Surveillance Capitalism: The Fight for the Future at the New Frontier of Power*. London: Profile Books.

Zuckerman, S. (1966) *Scientists at War*. London: Hamish Hamilton.

Index